Tomato Cultivation and Consumption - Innovation and Sustainability

Edited by Francesco Lops

Published in London, United Kingdom

Tomato Cultivation and Consumption - Innovation and Sustainability
http://dx.doi.org/10.5772/intechopen.111285
Edited by Francesco Lops

Contributors
Patta Sujatha, Madagoni Madhavi, Yarasi Bharathi, Mandalapu Pallavi, Polneni Jagan Mohan Rao, Bodduluru Rajeswari, Saddy Praveen Kumar, Anumala Akhil Reddy, Mohamed Youssef Banora, Owais Iqbal, Chengyun Li, Abdul Mubeen Lodhi, Nasir Ahmed Rajput, Raquel Saraiva, Igor Dias, José Grego, Margarida Oliveira, Jithesh Mundaya Narayanan, Tulasi Tirumalai Ramanujam, Kruthika Nagendra Rao, Vishwini Viswanathan, Albys Esther Ferrer Dubois, Yilan Fung Boix, Clara Esther Martínez Manrique, Liliana María Gómez Luna, Elizabeth Isaac Aleman, Sophie Hendrix Hendrix, Natalie Beenaerts, Ann Cuypers

First published in London, United Kingdom, 2024 by IntechOpen
IntechOpen is the global imprint of INTECHOPEN LIMITED, registered in England and Wales, registration number: 11086078, 5 Princes Gate Court, London, SW7 2QJ, United Kingdom

British Library Cataloguing-in-Publication Data
A catalogue record for this book is available from the British Library

Additional hard and PDF copies can be obtained from orders@intechopen.com

Tomato Cultivation and Consumption - Innovation and Sustainability
Edited by Francesco Lops
p. cm.
Print ISBN 978-1-83768-292-8
Online ISBN 978-1-83768-293-5
eBook (PDF) ISBN 978-1-83768-294-2

For EU product safety concerns:
IN TECH d.o.o., Prolaz Marije Krucifikse Kozulić 3, 51000 Rijeka, Croatia,
info@intechopen.com or visit our website at intechopen.com.

Meet the editor

Francesco Lops obtained a Ph.D. in Forest Plant Pathology from the Institute of Plant Pathology, University of Bologna, Italy, with research on chromogenic fungi's role in pine forest decay conducted at the Consortium Center, University of Basilicata, Italy. He held a National Research Council scholarship on fungal disease transmission and plant defense at the University of Turin, Italy, from 1995 to 1996. Dr. Lops was a post-doctoral scholar at the University of Bari, Italy from 1996 to 1997. In 2001, he won a university research competition. He became Associate Professor in Plant Pathology at the University of Foggia, Italy, in 2006. He contributes to the regional Xylella emergency task force in Puglia, Italy.

Contents

Preface

This collection of essays provides a deep dive into the intricate dynamics that define tomato cultivation, a plant that extends beyond its culinary role to become a crucial subject of study in the fields of agronomy and biology. Divided into two distinct sections - "Biological Interactions" and "Agronomic Practices" - the work presents a detailed analysis of the challenges and innovations that characterize the complex landscape of pomological research.

The first section, "Biological Interactions", introduces us to the realm of beneficial microorganisms with a particular focus on the Bacillus genus. Chapter 1 explores these Gram-positive bacteria's potential in the biological control of phytopathogens. From the direct suppression of microbial germination to the stimulation of plant development, the Bacillus genus emerges as a multifunctional agent. The detailed analysis reveals the production of antifungal compounds that play a crucial role in preventing and controlling plant diseases. This section provides a comprehensive framework of the strategies adopted by Bacillus, with special emphasis on its role as a plant growth-promoting rhizobacterium.

Chapter 2 focuses on the impact of root-knot nematodes on tomatoes. Examining current management strategies, the work highlights the importance of addressing these pathogens sustainably. Through the analysis of molecular interactions between tomatoes and nematodes, the work offers insights into future strategies to enhance plant resistance.

Finally, we conclude this section by exploring innovations in biological seed coatings. Delving into the rapidly growing world of this practice in the global seed market, Chapter 3 highlights how the application of beneficial microorganisms to seeds can provide protective coverage, improve shelf life, and promote healthier plant growth. This section reveals how biological practices can offer sustainable alternatives to traditional chemical methodologies.

The second section, "Agronomic Practices", outlines innovative approaches to tomato crop management. Chapter 4 focuses on advanced technologies in greenhouse cultivation and their impact on the Mediterranean region. Examining technological developments and trends in the use of treated wastewater, the work projects a future where sustainability merges with productivity, crucial to addressing the effects of climate change.

Chapter 5 delves into environmental challenges, providing innovative strategies to improve tomato productivity under stressful conditions. Using techniques such as "seed priming" and "mechanical conditioning," the work offers practical perspectives to overcome environmental obstacles and enhance crop resilience.

Finally, we conclude this section with a glimpse into the future of tomato crop management through the use of static magnetic fields in irrigation. Examining the effect on minerals and nutrients, Chapter 6 emphasizes how this technology can improve plant nutrition and increase the nutraceutical value of fruits.

In conclusion, this collection of research works not only provides a comprehensive overview of the world of tomato cultivation but also offers practical solutions to address current and future challenges. We hope this preface and the subsequent works are a valuable resource for those seeking significant advancements in sustainable tomato production.

Francesco Lops
Department of Agriculture, Food, Natural Resources and Engineering (DAFNE),
University of Foggia,
Foggia, Italy

Section 1

Biological Interactions

Chapter 1

Management of Phytopathogens by Antagonistic *Bacillus spp* in Tomato Crop

Owais Iqbal, Chengyun Li, Nasir Ahmed Rajput and Abdul Mubeen Lodhi

Abstract

Bacillus is a genus of gram-positive bacteria that is widely distributed in the environment. The species of this genus present in the endosphere, phyllosphere and rhizosphere in the plant and perform as a beneficial biocontrol agent and promote plant health. These strains exhibit diverse capabilities, including the potentiality to directly suppress the germination of microbial, stimulate plant development, reduce pathogen infections, degrade different types of hydrocarbons, function effectively across a wide temperature range, and induce immune resistance in host plants. The species/strains of *Bacillus* genus have proven promising biocontrol agents against a large number of fungal and bacterial causal organism, as well as plant-damaging insects. They induce a wide range of composites with antifungal properties, such as iturin, surfactin, cyclic lipopeptides, bacillomycin, bacteriocins, polyketide, lentibiotics, phospholipid, polyketide microlectine, isocosmarin and amino sugar. These compounds play a crucial role in preventing and controlling diseases in plants. The synthesis of these compounds is initiated in response to the presence of bacterial and fungal pathogen biomass and their cell walls. The purpose of this review is to offer a thorough exploration of the disease suppression mechanisms utilized by *Bacillus*, with a specific emphasis on their function as plant growth-promoting rhizobacteria (PGPR).

Keywords: tomato, diseases, phytopathogens, biocontrol, *Bacillus*

1. Introduction

Tomato (*Solanum lycopersicum* L.) is a highly significant and economically valuable agricultural crop cultivated worldwide, in both field and greenhouse conditions [1]. Tomatoes hold a prominent position in the agricultural sector, with a cultivated area spanning over 5 million hectares. This substantial acreage places tomatoes as the second most widely grown crop after potatoes. The global production of tomatoes is truly remarkable, exceeding an astounding 182 million tons [2, 3]. Despite its widespread cultivation, tomato crops are vulnerable to a range of challenges, including abiotic and biotic stresses [4]. Biotic stresses, arising from factors such as fungi, bacteria, phytoplasmas, viruses, and viroids, significantly impact crop productivity in both greenhouse and field environments [5]. Among these stresses, fungi present one of the most devastating problems, as they affect tomato plants with numerous fungal pathogens [6]. In fact, [7] demonstrated that fungal pathogens are responsible for more than 50% of diseases in tomato plants. Although various methods, such as soil solarization, fungicide seed dressing, spraying of fungicides and bactericides, crop rotation, field sanitation, and soil fumigation, are utilized to control plant diseases, their success is limited [8]. Moreover, treatments like soil fumigation do not provide long-term protection throughout the growing season. High infections often turnout in troublesome areas where soil fumigation was implemented prior the planting season, indicating contamination sources that can be traced back to the growers [9]. While chemical fungicides offer acceptable control of plant diseases in the field, excessive use of these fungicides has been reported to cause environmental pollution and have a destructive impact on human health [10]. In addition, biocontrol management of plant diseases extend a promising and surrogate method to hazardous chemical fungicides for controlling various plant diseases in both field and greenhouse settings. The growing interest in this emerging field can be attributed to a widespread desire to decrease dependence on agrochemicals, owing to their adverse impacts on human health and the environment. As a result, there is an increasing focus on alternative methods and strategies that promote sustainable and environmentally-friendly agricultural practices. The well-known main fungal and bacterial bio-control agents, such as *Trichoderma* spp., *Paecilomyces* spp., *Bacillus* spp., and *Pseudomonas* spp., have exhibited remarkable abilities in managing a wide range of plant diseases while also promoting plant growth [11]. Moreover, these biological control agents (BCAs) possess a host of additional favorable traits, including rhizosphere competence, fungicide tolerance, saprophytic competitiveness, temperature tolerance, edaphic adaptability, beneficial searching potential, host selectivity, increased reproduction rate, short life cycle, adaptability, and the ability to persist even after reducing the host population [12].

Out of these fungal and bacterial genera, *Bacillus* is the most well-known and extensively studied bacterial genus. Since the late 1800s, the study of *Bacilli* has encompassed traditional microbiology and biochemistry methods, as well as more advanced techniques like genomic and proteomic analysis [13]. *Bacillus* refers to a type of bacterium that is rod-shaped, gram-positive, and can exist in either aerobic or facultative anaerobic conditions [14]. In response to various environmental or nutritional pressures, *Bacillus* can produce highly resilient endospores that remain dormant for extended periods [15, 16]. Numerous *Bacillus* species have revealed their role in enhancing plant growth parameters and have proven to be promising biocontrol agents against various plant diseases. Furthermore, they play a role in enhancing plant resilience to both abiotic and biotic stress factors [17, 18]. Several studies have shown that species within this genus promote plant growth through the production of antibiotics, phytohormones, lipopeptides, antimicrobial compounds, nutrient acquirement (such as N and P), and endospores formation, leading to a longer shelf life as well as enhanced plant growth [19–22]. Specific strains of various *Bacillus* species have proven effective in controlling fungal pathogens, including *Fusarium*, *Rhizoctonia*, *Oidium*, *Septoria, Macrophomina*, *Botrytis*, *Pythium*, *Verticillium*, *Phytophthora*, *Sclerotium*, and *Alternaria*. They have also demonstrated efficacy against bacterial phytopathogens, including *Erwinia*, *Pseudomonas syringae*, *Ralstonia*, and *Xanthomonas* (**Figure 1**). Notably, certain strains of *B. cereus*, *B. subtilis*, *B. megaterium*, *B. velezensis*, *B. amyloliquefaciens*, and other *Bacillus* species have revealed high effectiveness in controlling numerous plant diseases, such as Fusarium wilt,

Figure 1.
Antibiotic compounds produced by Bacillus against fungal and bacterial phytopathogens.

damping off, grey mold, crown, powdery mildew, verticillium wilt, late and early blight, septoria leaf spot, bacterial wilt, bacterial soft rot, bacterial spot, bacterial speck, and various foliage diseases (**Table 1**). Therefore, this review aims to explore the promising biocontrol abilities of different *Bacillus* species/strains against major groups of fungal and bacterial pathogens.

2. Diseases, symptoms, causal organism and nature of pathogen in tomato crop

Tomato is the most popular vegetable crop cultivated worldwide [70]. It serves as an excellent source of nutrition, containing essential components such as vitamin C, potassium, carotenoids, and various phytochemical compounds [77, 78]. However, crops belonging to the *Solanaceae* family, including tomato, are highly susceptible to various fungal diseases that significantly reduce both the quality and yield of the crop [63]. These diseases like damping off, grey mold, fusarium wilt, powdery mildew, crown, fruit and root rot, verticillium wilt, late and early blight, bacterial soft rot, septoria leaf spot bacterial wilt, bacterial spot, and bacterial speck, can cause yield losses ranging from 10 to 90% [3, 29, 37, 38, 47, 61].

Diseases	Causal organism	Species	Reference
Bacterial wilt	*Ralstonia solanacearum, Fusarium oxysporum f. sp. radicis-lycopersici, Ralstonia Pseudosolanacearum,*	*Bacillus Amyloliquefaciens, B. subtilis, B. cereus, B. pumilus, B. licheniformis B. methylotrophicus, B. velezensis*	[8, 23–28]
Gray mold	*Botrytis cinerea*	*Bacillus cabrialesii, B. cereus, B. firmus, B. megaterium, B. endophyticus, B. aryabhattai, B. velezensis, B. subtilis, B. licheniformis, B. amyloliquefaciens, B. methylotrophicus, B. halotolerans*	[17, 29–36]
Fusarium wilt	*Fusarium solani, Fusarium oxysporum f. sp. lycopersici*	*Bacilllus megaterium, B. amyloliquefaciens, B. cereus, B. velezensis, B. methylotrophicus, B. huringiensis, B. pumilus, B. subtilis*	[9, 18, 37–44]
Bacterial soft rot	*Erwinia carotovora*	*B. firmus, B. megaterium, B. endophyticus, B. aryabhattai, B. velezensis, B. subtilis, Bacillus cereus*	[35, 45, 46]
Bacterial speck	*Pseudomonas syringae*	*Bacillus cereus, B. firmus, B. megaterium, B. endophyticus, B. aryabhattai, B. velezensis, B. subtilis*	[35, 47, 48]
Bacterial spot	*X. perforans, X. campestris, X. vesicatoria* and *X. gardneri, Xanthomonas euvesicatoria*	*B. subtilis, B. amyloliquefaciens B. velezensis, B. pumilus, B. cereus, B. methylotrophicus*	[49–54]
Damping-off	*Rhizoctonia solani, Pythium aphanidermatum, Fusarium solani, Fusarium oxysporium*	*Bacillus subtilis, B. firmus, B. megaterium, B. endophyticus, B. aryabhattai, B. velezensis, B. polymyxa, B. thracis, B. circulans* and *B. polymyxa and B. sphaericus, B. amyloliquefaciens*	[16, 35, 55–60]
Verticillium wilt	*Verticillium dahliae & V. longisporum*	*B. amyloliquefaciens, B. firmus, B. megaterium, B. endophyticus, B. aryabhattai, B. velezensis, B. subtilis, B. thuringiensis* and *B. weihenstephanensis, B. cereus*	[30, 42, 61, 62]
Late blight	*Phytophthora infestans*	*B. subtilis, B. firmus, B. megaterium, B. endophyticus, B. aryabhattai, B. velezensis, B. pumilus, B. cereus*	[35, 63–67]
early blight	*Alternaria solani & Alterneria alternata*	*B. amyloliquefaciens, B. cereus, B. firmus, B. megaterium, B. endophyticus, B. aryabhattai, B. velezensis, B. subtilis*	[64–66, 68, 69]
Septoria leaf spot	*Septoria lycopersici*	*B. cereus, B. amyloliquefaciens, B. subtilis, B. thuringiensis*	[65, 66, 68, 70]
Powdery mildew	*Oidium lycopersicum & Oidium neolycopersici*	*Bacillus subtilis, B. amyloliquefaciens*	[65, 71, 72]
Crown and root rot	*Pseudomonas solanacearum, R. solani, Phytophthora capsici, F. oxysporum f. sp. lycopersici & radicis-lycopersici (Forl)*	*B. subtilis, B. pumilus, B. siamensis, B. thuringiensis, B. amyloliquefaciens*	[1, 26, 73–76]

Table 1.
The antagonistic effect of Bacillus *sp., for controlling of various diseases in tomato crop caused by phytopathogens.*

2.1 Crown and root rot disease

F. oxysporum f. sp. radicis-lycopersici (FORL) is the causal agent of crown and root rot disease in tomato crops. This disease for first time in 1974 was reported in Japan and has since become widespread in various regions across the world [79, 80]. It stands as a significant soil-borne ailment affecting tomato cultivation [73]. The fungus exhibits the capability to infect both tomato seedlings within transplant houses and fully grown plants in the open field [1]. *F. oxysporum* severely infects the root and collar of a plant during the early stages' growth, leading to decreased productivity in field and greenhouse conditions [81]. The symptoms exhibited by affected tomato plants include stunted growth, leaf chlorosis and premature shedding lower true leaves. As the disease progresses, a distinct brown lesion develops around the root and shoot junction, by showing root rot, wilting, and leading toward plant death [82]. In both greenhouse and field environments, the disease has been observed to result in crop yield reductions spanning from 20 to 60% [83].

2.2 Damping-off

Tomato damping-off, which is triggered by pathogens like *Pythium aphanidermatum* and *Rhizoctonia solani* [55], *F. solani* [56], and *F. oxysporum* [16, 57], is one of the most aggressive and destructive diseases of tomato crop. This disease leads to the deterioration of germinating seeds and juvenile seedlings, resulting in a significant reduction in crop productivity. Damping-off poses a major problem for cultivators, whether they are operating in greenhouses or in the field [84]. In the literature, the definition of damping-off is challenging due to varying perspectives among researchers. Some consider it a sickness, while others view it as a symptomatic condition [85, 86]. Damping-off can be categorized into two types: pre-emergence and post-emergence. Soil-borne pathogens are commonly responsible for the spread of damping-off, and play a major role in post emergence of the disease [87]. The disease symptoms of damping-off in tomato plants manifest in two distinct ways. Firstly, after emergence, the seedling's hypocotyl becomes discolored and water-soaked at the soil surface, ultimately leading to plant death [88]. Secondly, under favorable conditions, the typical symptoms include stunted growth, the presence of brown and small water-soaked lesions around the leaves, and infection of the entire root system. Often, seedlings die before they can even emerge [89]. In the field, losses of up to 100% have been recorded [90].

2.3 Fusarium wilts diseases

Fusarium wilt disease of tomatoes, caused by Fusarium species such as *F. oxysporum f. sp. lycopersici* and *F. solani*, is a highly destructive constraint in tomato production [18, 39, 40]. The disease was first reported by W.C. Snyder and H.N. Hansen in 1940 [91]. Fusarium species can induce a fatal vascular disease known as wilt, primarily affecting plants of the *Solanaceae* family [92]. This soil-borne fungal disease significantly impacts tomato crop yield in regions like tropical and subtropical, as the pathogen has the potential to persist in the soil medium for maximum periods of time and infiltrate plants through root injuries [38]. The aggressive nature of this disease presents a significant challenge in terms of control, primarily due to the pathogen's ability to attack vascular tissue and its soil-borne nature [39]. In the presence of the pathogen, the disease manifests as obstruction and degradation of the xylem within the host plant [93]. Typical symptoms include vascular wilting, yellowing, leaves' wilting, and discoloration of the vascular tissue, which may turn dark brown. The disease may exhibit symptoms like stunted growth and may eventually lead to the death of the entire plant. In greenhouse and field condition, it was estimated that yield losses may be in range of 50% due to Fusarium wilt [1].

2.4 Verticillium wilt

Verticillium wilt, also known as vascular wilt, poses a serious challenge to sustainable development and cultivation of tomato plant. A number of species of Verticillium fungus such as *V. dahliae* and *V. longisporum*, has been reported and that may leads decline yield production in both yield and fruit quality [30, 61]. Verticillium wilt in tomatoes was reported in 1928 by H.C. Pierce and W.C. Snyder [94]. Unfortunately, resistant tomato varieties were not available at that time to combat this destructive disease [62]. The Verticillium fungus obstructing vascular system of a plant and lead to wilt and death of plant consequently, and the affected plants exhibit symptoms such as wilting [95]. By infiltrating the tomato plant via the roots, this fungus establishes itself within the xylem vessels and generates adhesive substances that hinder the transportation of water and nutrients [96, 97]. Furthermore, it produces harmful compounds that stimulate the plant's defense mechanism, resulting in the generation of reactive oxygen species (ROS) and the reinforcement of the cell wall [98]. Verticillium wilt disease of tomato crops symptoms includes yellowing and curling of the leaves, particularly at the lower portion of the plants, as well as stunted growth and browning or discoloration of the stems, especially near the plant's base. Dark brown streaks in the vascular tissue may also be present, indicating damage to the plant's vascular system [89, 99]. Yield losses attributed to verticillium wilt in tomato crops can range from 10 to 50% or even more [100]. Consequently, effective strategies and resistant varieties are needed to combat this devastating disease and protect tomato crop productivity.

2.5 Grey mold

A variety of diseases pose a threat to tomato crops, and among them is gray mold, also referred to as Botrytis gray mold, and caused by *Botrytis cinerea* airborne necrotrophic fungal pathogen which has impact on the quality and production of tomato globally [29, 30]. The discovery of gray mold affecting tomatoes was initially reported by Whetzel and Hesler in 1923 [101]. Particularly vulnerable to gray mold are fresh-market tomatoes, making it a crucial concern both before and after harvest [31]. The pathogen attacks different parts of the tomato plant including, flowers, leaves, stems and fruits, leading to significant damage [102]. Under cool and humid weather conditions, the fungus flourishes, leading to the formation of a grayish mold on the affected areas of the plant. *B. cinerea* affects approximately 200 plant species and causes lesions, producing numerous spores on the above-ground sections of plants [103]. Gray mold disease presents distinct symptoms, such as the presence of grayish-brown fuzzy mold on the leaves surface, stems and tomato fruit. Infected leaves and stems may also exhibit a covering of grayish-brown fuzzy mold, while turning brown or black in color. The fruit manifests grayish-brown mold symptoms and may become soft and mushy. Plants affected by this disease experience stunted growth and a decline in fruit quality and yield, whether in greenhouses or fields [29, 32–34]. Gray mold causes significant losses, ranging from 5 to 20% of the crop in greenhouse and field environment [104].

2.6 Late blight disease

Late blight, a destructive disease affecting tomatoes, is caused by *Phytophthora infestans* (Mont.) de Bary, which belongs to the family Oomycetes. Oomycetes are a unique group of filamentous eukaryotes [105]. The pathogen devastating impacts on several plant species, comprising those in the *Solanaceae* family. During the nineteenth century, the pathogen caused widespread devastation to potato crops, leading to the catastrophic event known as the Irish potato famine [106]. Unfortunately, during that time, no resistant cultivars were available for tomatoes or potatoes to combat late blight disease [64]. This pathogen thrives in cool and humid weather conditions and produce visible symptoms on entire parts of the plant at any stage of growth in form of dark lesions [107]. It causes extensive damage to the stems, leaves, and fruits, often leading to complete crop losses within a mere two weeks [108]. The characteristic symptoms of late blight manifest as water-soaked lesion on entire plant parts, which initially appear light green and later turn brown or black [109]. As the disease progresses, it spreads rapidly, transforming into white mold symptoms on the undersides of stems and leaves. Affected leaves develop irregular shapes, accompanied by dark brown lesions which are often encircled by a yellow halo [110]. Additionally, the fruit can start rotting at the stem end or any other part. Infected plants exhibit symptoms of yellowing, wilting, and eventual death [89]. This devastating disease inflicts annual losses exceeding five billion USD on the crop, as reported by various sources [111].

2.7 Early blight disease

Early blight is one the most important fungal disease caused by *Alterneria solani* (Ellis and Martin) Jones and Grout and *A. alternata*, that poses a serious threat to tomato crops production [65, 68]. This fungal pathogen is not limited to tomatoes but also affecting other plants like potatoes and eggplants [112]. *A. solani* a soil-borne as well as airborne pathogen responsible for diseases like leaf blight, collar, and fruit rot in tomatoes, whereas the disease can be spread through fungal spores [113]. It was first reported in early nineteenth centuries in the US, quickly becoming a significant concern for tomato growers [114]. In regions with high levels of dew, rainfall, humidity, and temperatures, can lead to complete defoliation of the leaves and particularly detrimental to tomato plants [115]. The pathogen produces enzymes like cellulases, which dissolve the cell wall of host's plant by pectin methyl galacturonase, which aids in host colonization. As a result, the disease negatively impacts crop production by causing premature defoliation, may become cause of low quality and quantity of fruit [116]. Initial symptoms of the disease include brown to black lesion appears on the lower leaves during the early stages of disease development, and gradually spreading to the upper parts of the plant over time. These spots typically are in range of 1/8 to 1/4 inch in diameter. In most cases, the disease manifests on the stem with dark and brown discoloration. Subsequently, the leaves exhibit yellowing symptoms, ultimately resulting in plant defoliation. Furthermore, early blight also affects the quality of tomato fruit, causing dark, sunken lesions [117]. The yield losses due to the disease have been reported in a range of 79% in tomato crops [118].

2.8 Powdery mildew disease

Powdery mildew disease of tomato caused by *Oidium lycopersicum* and *O. neolycopersici* are the major constrains in tomato production that affects all parts of the tomato plant

[71, 72, 119]. This disease was observed for first time in England in 1986 and 1987, and it has then spread world-wide [120, 121]. Consequently, the new mildew pathogen on tomato plants was variously termed *O. lycopersicum, Erysiphe orontii* or *E. cichoracearum*) or was simply described as *Erysiphe sp.*. The first appropriate description of the fungus, *Oidium lycopersicum*, appeared to come from Australia, and the name was re-designated, in 1999, as *Oidium lycopersici*, in accordance with the International Code of Botanical Literature [122]. However, confusion remained over classification based on morphological characteristics. Consequently, we analysed the internal transcribed spacer regions of the nuclear rRNA genes from the new tomato powdery mildew pathogen and were able to differentiate *Oidium (neo)lycopersici* from *E. orontii* and *E. cichoracearum*. Moreover, we found *O. (neo) lycopersici* to be a sister taxon of *E. aquilegia var. ranunculi*. Importantly, it was recognized that all recent outbreaks of tomato powdery mildew reported outside Australia were caused by a species that formed conidia singly, or, in high relative humidity, in pseudo-chains of 2–6 conidia, and so created a new species, *O. neolycopersici*, for this pathogen. The Australian isolates, which always formed conidia in chains, retained the name *Oidium lycopersici*. However, its true identity was uncertain due to the lack of a sexual stage and varying reports of its structure, particularly whether conidia were formed singly or in chains. *O. lycopersicum* (Erysiphales) differs from *L. taurica* based on a number of characteristics, including conidiophore and conidia morphology, and from *E. cichoracearum*, which produces conidia in long chains [123]. It has the ability to infect the crop at any stage of growth in a greenhouse or in the field, leading to a reduction in fruit quality and yield. Consequently, a huge number of tomato cultivars are susceptible to *Oidium* spp. [124]. This disease can manifest at any stage of the crop by producing symptoms likes spots on the leaves, stems, and fruits, and may lead to cover the entire plant parts [119]. Subsequently, the leaves become distorted and stunted, exhibiting curling and twisting [125]. These dark spots or blotches appeared on the infected fruits, damaging their shape and color of the fruit and ultimately lead to loss of fruit in the form of its quality. In most cases, premature leaf drop occurs, leading to a decrease in photosynthesis that contributes to yield reduction [126] and the estimated yield was around 50% in fruit and yield, in control environment as well as in field conditions [127].

2.9 Septoria leaf spot disease

Septoria leaf spot (also named as septoria blight) is caused by *Septoria lycopersici*, most aggressive and destructive pathogen of vegetables. It contributes huge losses to tomato production in market and poses a major threat to tomato production worldwide [66]. The earliest report of Septoria leaf spot in the United States came from Byron D. Halsted, an American plant pathologist, who observed it in New Jersey during the years 1894 to 1895. Since then, the disease has spread to tomato-growing areas worldwide, presenting continuous challenges for tomato growers [128]. Disease incidence tends to increase significantly during the summer when temperatures reach their peak and precipitation is high. When temperatures exceed 25°C and leaves remain wet for extended periods, tomato yields can decrease by more than 50% [129]. The pathogen primarily spreads through contaminated seeds, but it can also survive in crop debris for extended periods [130]. Septoria leaf spot affects plants at any stage when temperatures range between 20 and 25°C, combined with high humidity and rain showers [131]. Under such conditions, the fungus initially targets older leaves, causing circular spots with dark brown margins and tan to gray centers. These spots are accompanied by black pycnidia, leading to extensive damage across the entire leaf area [132, 133]. The disease also manifests as small, dark lesions measuring approximately 1–8 mm in diameter on stems, peduncles, and calyxes [134]. This ultimately results in a significant reduction in fruit yield, not only due to the loss of photosynthetic area but also because the fruit becomes more susceptible to sunburn. In severe cases, tomato crops have experienced estimated disease losses of up to 100% [135].

2.10 Leaf mould

Leaf mould of tomato caused by Cladosporium fulvum syn. *Passalora fulva*, is one of the most devastating diseases in tomato crop [80]. The disease was first time reported in Netherlands [136]. Typically, the fungus primarily affects the foliage of tomato plants, although there are instances where stems, blossoms, petioles, and fruit can also be targeted. Successful infection occurs when the fungus conidia settle on the lower side of a leaf, germinate, and enter the plant through open stomata [137]. Symptoms of the disease become apparent approximately one week after infection, manifesting as pale green or yellowish diffuse spots on the upper surface of the leaves. Over time, these spots enlarge and become distinctive yellow patches due to cell death in the palisade parenchyma. The most noticeable symptoms are observed on the lower side of the leaf, where patches of white to olive-green mould develop, eventually turning brown when sporulation begins [138]. In the initially stage of the disease, the stomata become blocked by clusters of conidiophores, which use the stomata to exit the leaf and release conidia, further contributing to the spread of the disease. Stomatal blockage severely impedes plant respiration, leading to leaf wilting, partial defoliation, and, in severe infections, the death of the affected plant [139]. In the favorable condition the disease cause up to 50% yield losses [140].

2.11 Bacterial soft rot disease

The presence of bacterial soft rot disease poses serious threat to tomato crops production, resulting in substantial yield losses in both greenhouse and open field environments, surpassing the impact of other bacterial diseases [35]. The initial occurrence of this disease was observed on tomato fruit in a greenhouse in the Buenos Aires province in 1995 [45]. The pathogen responsible for this disease is known as *Erwinia carotovora* subsp. carotovora (Ecc) [141]. *Ecc*, a rod-shaped bacterium, affects a wide range of vegetable crops, including potatoes, carrots, onions, cucumbers, lettuce, and tomatoes. It can also infect ornamental plants under favorable environmental conditions [142]. The pathogen has the ability to infect all parts of the plant and exhibits severe symptoms in tomato plants. Initially, yellowing symptoms on the lower side of leaves, led by brown-yellowing of the pith and stem xylem vessel [143]. As the disease progresses, the entire tomato plant wilts, showing water-soaked lesion all over the stem, vascular tissue turn brown, pith become hollow, and rotting of stems and fruits. The symptoms typically begin in the root or crown region of seedlings in greenhouses or fields [144]. It has been reported that yield losses up to 100% may be reached due to the bacterial soft rot disease [145].

2.12 Bacterial speck

Pseudomonas syringae (Okabe) Alstatt is the causal agent of Bacterial speck of tomato, is the most prevalent bacterial disease infecting tomato plants worldwide [146]. The first recorded instance the disease in tomato plants was reported by J.C.

Walker in New York in 1922 [147]. Since then, the disease has been spread to rest of the world and identified in various tomato-growing regions worldwide and continues to pose a significant threat to tomato production, whether in greenhouses or open fields. The pathogen is commonly introduced through infected seeds or may already be present in the surrounding environment [148]. The disease manifests as small black spots with a sunken appearance on green tomato fruits, accompanied by darker green halos that eventually develop into cankers. Ripe tomato fruits exhibit dark brown to black spots, ranging from 1 to 2 mm in diameter. Additionally, bacterial speck disease often presents with large black spots on older leaves, stems, and petioles [149, 150]. These symptoms contribute to various losses, including reduced fruit yield and quality, diminished plant vigor, and heightened susceptibility to other diseases [151]. In severe cases, bacterial speck can lead to defoliation and even death of the plant. The extent of these losses depends on the severity of the infection, the susceptibility of the tomato cultivar, and environmental factors such as temperature and moisture [152, 153]. It is estimated that bacterial speck disease causes a staggering 52% loss in tomato crops [154]. Given its detrimental impact, effective management strategies are crucial to mitigate the economic and agricultural consequences associated with this disease.

2.13 Bacterial wilt

Bacterial wilt of tomato, caused by *Ralstonia pseudosolanacearum* and *R. solanacearum* [155], are the major constrain in tomato losses in temperate, tropical and subtropical regions in the worldwide [23]. These pathogens have the ability to infect around 200 plant species, including those in the *Solanaceae* family [24]. The typical symptoms exhibited an infected tomato plants may include stunted growth, yellowing of leaves, vascular discoloration, wilting and ultimate cause death, all of which are associated with the presence of the bacterial wilt pathogens *R. solanacearum* and *R. pseudosolanacearum* [25]. When stems affected by the disease are submerged in water, bacteria continuously seep out from the cut ends. Transmission of *R. solanacearum* occurs through various means, including root contact, irrigation water, insects, and machinery [8, 26, 71]. In tomato crops, this disease can result in yield losses of up to 91% [156].

2.14 Bacterial spot disease

Bacterial spot of tomato is a significant foliar disease that affects the plant at any stage and hampers the growth of both fruit and plant [49]. The disease is caused by five species of the bacterial genus *Xanthomonas*: *X. euvesicatoria*, *X. perforans*, *X. campestris*, *X. vesicatoria*, and *X. gardneri*. These bacteria result in substantial losses in tomato production [49–54]. This disease was first observed in tomato plants in the United States in the early 1900s, specifically in the southern states of Georgia and South Carolina [157]. The pathogen responsible for this disease can spread through contaminated seeds, live as epiphytes on tomato leaves, and infect various parts of the plant by inducing spots, lesions, and defoliation. It can also persist on weeds and infected plant debris [158, 159]. Furthermore, heavy rainfall, contaminated mechanical tools, and insects serve as additional vectors for spreading the pathogen [160]. The pathogen produces numerous bacteriocin-like substances that pose a risk to fruit and plant growth [161]. Due to heavy rainfall, the disease exhibits early symptoms on the leaves, stem, and fruits, thereby reducing the quality and yield of the fruit [125]. As the disease progresses, small, water-soaked, round dark brown spots appear on the affected parts of the leaves, later turning black with a near-yellow halo [162]. These water-soaked spots increase in size and undergo a color transformation from dark green to purplish-gray, accompanied by the emergence of a recognizable black center. Over time, the leaf spots expand and merge, creating a scorching appearance [163]. This destructive disease has been recorded to cause losses of up to 50% in greenhouse tomato crops and overall yield [164].

3. Control of fungal and bacterial disease through *Bacillus* sp., in tomato crop

Chemical fungicides are commonly utilized to prevent various pests and diseases, but they have limited effectiveness and pose harm to crop physiology, human health and also harmful impact on environment [1]. In order to manage these diseases and decrease pathogen populations in crops, novel and alternative approaches like the utilization of fungi and bacteria as biocontrol agents have been implemented [24]. The possession of both antagonistic and plant growth-promoting characteristics by biocontrol agents is considered significant for managing plant diseases and improving fruit yield. Numerous rhizobacteria including *Bacillus*, *Streptomyces*, *Pseudomonas*, *Flavobacterium*, *Brevibacillus*, *Mesorhizobium* and *Rhizobium* found in plants exhibit growth-promoting properties, capable of mitigating pathogen infections and enhancing plant development [165]. Among them, *Bacillus* sp., are well-known for their superior bacterial antagonistic properties compared to other genera because they can survive at higher temperatures and resist desiccation by producing endospores within the cell. Additionally, *Bacillus* sp., can promote plant growth to some extent under unfavorable conditions [166]. *Bacillus* species possess the capability to decrease populations of fungal pathogens through the production of a substantial quantity of antibiotics. They achieve this by lysing cells, promoting plant growth, and triggering resistance against a range of diseases. *Bacillus* species are capable of producing a variety of antibiotics, including iturin, surfactin, cyclic lipopeptides, bacillomycin, bacteriocins, polyketide, lentibiotics, phospholipid, polyketide microlectine, isocosmarin and amino sugar. These antibiotics play a crucial role in inhibiting the growth and spread of bacterial and fungal pathogens, which are responsible for causing diseases in plants. By producing these diverse compounds, Bacillus species effectively suppress the harmful microorganisms, thus protecting the health and vitality of plants. Out of the 21 isolates tested, only *B. amyloliquefaciens* (FZB24) showed maximum growth inhibition of *F. oxysporum* through dual assay test. The study also revealed that *B. amyloliquefaciens* was further efficacious in reducing the occurrence of Fusarium wilt disease in tomato plants due to its abundance of antibiotics and enzymes, including polyphenol oxidase, ammonia lyase, phenylalanine catalase, superoxide dismutase and peroxidase [41]. *Bacillus velezensis*, recover from the crown root tissue of tomatoes, produced various compounds such as lipopeptides, polyketides, dipeptide bacilysin, and volatile substances including fengycin, bacillomycin, surfactin, benzaldehyde, bacillaene, macrolactin, difficidin, tetradecane, dipeptide bacilysin, benzeneacetic acid, benzaldehyde, phenylethyl alcohol and 1-decene. Additionally, the specie exhibited the ability to generate lytic enzymes (chitinase, protease, and β-glucanase), indole-3-acetic acid, solubilize inorganic phosphate and siderophore. *B. velezensis* achieved a maximum growth suppression on the mycelial development of *V. dahliae* and decrease the prevalence of wilt disease in tomatoes by 70.43% in both greenhouse and open field conditions [42]. Two selected isolates of *Bacillus* produced antibiotics, cyanide, and solubilized phosphate, resulting in a 44% growth inhibition against *F. oxysporum*, responsible for the root and crown rot disease in tomatoes [73]. Due to the presence of polyphenol oxidase, superoxide, catalase, dismutase, and peroxidase activities produced by *Bacillus subtilis* CBR05 the treated tomato plants exhibited a minimum occurrence (36%) of soft rot disease [46]. In another study, *B. subtilis* suppress the root and crown rot disease incidence by up to 75% in a greenhouse [9]. A total of 200 different strains of *Bacillus* were obtained from the rhizosphere soil of tomatoes and potatoes. These strains were carefully examined and tested to determine their ability to antagonist or inhibit the growth and activity of the bacterial wilt pathogen *Ralstonia solanacearum*. Out of these strains, only four strains showed promising results in combating bacterial wilt disease. Specifically, two strains of *B. amyloliquefaciens* (AM1 and D29), one strain of *B. subtilis* (D16), and one strain of *B. methylotrophicus* (H8) displayed significant decreases in the prevalence of bacterial wilt disease. The reduction in disease incidence ranged from 81.1 to 89.0%, indicating a strong antagonistic effect against the pathogen *Ralstonia solanacearum*. Furthermore, these four strains demonstrated additional beneficial effects on the growth and development of the plants. They were found to enhance plant height by up to 20–45% and also increase the dry weight of plant by up to 45–92%. These positive effects were attributed to the production of certain compounds such as indole-3-acetic acid, phosphate solubilization and siderophores by the *Bacillus* strains. These compounds are known to promote plant growth and nutrient uptake. The greenhouse experiments revealed the potential of these specific strains of *Bacillus* in improving plant health and productivity, providing a promising solution for managing bacterial wilt disease in tomato and potato crops [8]. Lamsal et al. [167] obtained seven *Bacillus* strains from the rhizosphere of tomato crops. These isolated strains exhibited a significant inhibitory effect, with more than 80% suppression of mycelial growth of the targeted pathogen. Moreover, they demonstrated a remarkable reduction of 74% in the incidence of late blight disease in a greenhouse environment. Additionally, the Bacillus strains were found to positively influence plant growth. Furthermore, the researchers explored the potential of a *B. subtilis* formulation as a seed treatment method. They discovered that this treatment was particularly effective in combating damping off disease, which is caused by the pathogen *Pythium aphanidermatum* [56]. Hentriacontane and 2,4-di-tert-butylphenol are metabolic compounds of *B. cereus* (MH778713). These compounds have demonstrated strong inhibitory properties against various fungal plant pathogens, including *F. oxysporum* and *Colletotrichum orbiculare*, which cause fusarium wilt in tomatoes [18]. Im et al. [27] successfully isolated two metabolites, namely oxydifficidin and difficidin derivatives, from the strain methylotrophicus (DR-08). These compounds were proven to possess significant effectiveness against the bacterial wilt pathogen *R. solanacearum*. Through pot and field experiments, the strain DR-08 demonstrated its ability to effectively suppress the development of bacterial wilt in tomatoes, as well as bacterial leaf spot symptoms on peach and red pepper plants. The desired outcome was achieved using a concentration of 30%. These findings highlight the potential of DR-08 as a biocontrol agent for managing bacterial diseases in various crops, including tomatoes, peaches, and red peppers.

4. Conclusion

In this review, we emphasize the capacity of microbial antagonists to efficiently control infectious plant diseases originating from fungal and bacterial pathogens in tomato crops. Various species/strains of *Bacillus*, including *B. subtilis*, *B. cereus, B. megaterium, B. endophyticus, B. velezensis, B. amyloliquefaciens*, and *B. methylotrophicus*, have been identified as having favorable characteristics as plant growth-promoting rhizobacteria (PGPR) and biocontrol capabilities. The identification of potent *Bacillus* strains typically involves extensive sampling from the rhizosphere of the specific host plant, followed by dual assay method and in vivo testing against the target plant pathogen(s). In some instances, more than a hundred *Bacillus* strains have proved antagonistic antifungal activity against fungal and bacterial pathogens through in vitro methods, with certain strains exhibiting high growth inhibition. The effective *Bacillus* strains produce various compounds such as iturin, surfactin, cyclic lipopeptides, bacillomycin, bacteriocins, polyketide, lentibiotics, phospholipid, polyketide microlectine, isocosmarin and amino sugar. These compounds performance well to suppressing bacterial and fungal pathogens responsible for numerous plant diseases. Each Bacillus strain produces unique metabolites and compounds, resulting in variations in type and quantity. Consequently, these compounds display strain-specific effects in inhibiting the growth and infection of specific phytopathogens. Studies have shown that mutant strains lacking specific compounds are ineffective in controlling the targeted pathogens. Generally, Bacillus bacteria inhabit the root rhizosphere and endosphere of host plants, forming mutually beneficial relationships with them. These bacteria colonize the root rhizosphere and endosphere and provide several benefits to the host plant, including nutrient acquisition, disease suppression, and stress tolerance. Therefore, Bacillus species have been widely utilized for controlling various diseases in plants, including fungal soil-borne, root infecting, seed-borne, crown, fruit and root rot, Fusarium wilt, Verticillium wilt, Septoria leaf spot, damping-off, grey mold, late and early blight, and powdery mildew. Moreover, they effectively combat some bacterial diseases viz., bacterial wilt, bacterial speck, bacterial spot, and bacterial soft rot. The addition of Bacillus cell cultures or cultural filtrates containing effective compounds to the soil surrounding the roots, as well as seed treatment with these compounds, offer comparable benefits in preventing plant pathogen infestations. In some cases, more efficient disease control can be achieved by combining *Bacillus* with other bacterial antagonists (e.g., *Pseudomonas*) or fungal antagonists (e.g., *Trichoderma* spp.) or by using compatible fungicides. Overall, the appearance of *Bacillus* in the root rhizosphere and endosphere significantly influences the health and productivity of host plants.

Acknowledgements

I thank Professor Abdul Mubeen Lodhi and Professor Chengyun Li for their inspiration and help to carry out this work.

Conflict of interest

The authors declare that there is no conflict of interest.

Author details

Owais Iqbal[1,2]*, Chengyun Li[2], Nasir Ahmed Rajput[3] and Abdul Mubeen Lodhi[1]

1 Department of Plant Protection, Sindh Agriculture University Tandojam, Pakistan

2 Key Laboratory of Agro-Biodiversity and Pest Management of Education Ministry of China, Yunnan Agricultural University, Kunming, China

3 Department of Plant Pathology, University of Agriculture, Faisalabad, Pakistan

*Address all correspondence to: owais.iqbal.918@gmail.com

References

[1] Khalil MMR, Fierro-Coronado RA, Peñuelas-Rubio O, Villa-Lerma AG, Plascencia-Jatomea R, Félix-Gastélum R, et al. Rhizospheric bacteria as potential biocontrol agents against Fusarium wilt and crown and root rot diseases in tomato. Saudi Journal of Biological Sciences. 2021;**28**(12):7460-7471

[2] Panno S, Davino S, Caruso AG, Bertacca S, Crnogorac A, Mandić A, et al. A review of the most common and economically important diseases that undermine the cultivation of tomato crop in the Mediterranean basin. Agronomy. 2021;**11**(11):2188

[3] Caruso AG, Bertacca S, Parrella G, Rizzo R, Davino S, Panno S. Tomato brown rugose fruit virus: A pathogen that is changing the tomato production worldwide. The Annals of Applied Biology. 2022;**18**(3):258-274

[4] Susič N, Žibrat U, Širca S, Strajnar P, Razinger J, Knapič M, et al. Discrimination between abiotic and biotic drought stress in tomatoes using hyperspectral imaging. Sensors and Actuators B. Chemical. 2018;**273**:842-852

[5] Hashem AH, Abdelaziz AM, Attia MS, Salem SS. Selenium and nano-selenium-mediated biotic stress tolerance in plants. In: Selenium and Nano-Selenium in Environmental Stress Management and Crop Quality Improvement. Vol. 2022. New York: Springer, Cham; 2022. pp. 209-226

[6] Jayakumar V, Prasad GS, Kumar K, Sharma TVRS, Srivastava RC. Hand Book on Crop Pests and Diseases of Andaman and Nicobar Islands. CARI: Port Blair; 2010. p. 104

[7] Bonanomi G, Antignani V, Pane C, Scala F. Suppression of soilborne fungal diseases with organic amendments. Journal of Plant Pathology. 2007;**98**(3):311-324

[8] Almoneafy AA, Xie GL, Tian WX, Xu LH, Zhang GQ, Ibrahim M. Characterization and evaluation of *Bacillus* isolates for their potential plant growth and biocontrol activities against tomato bacterial wilt. African Journal of Biotechnology. 2012;**11**(28):7193-7201

[9] Baysal Ö, Çalışkan M, Yeşilova Ö. An inhibitory effect of a new Bacillus subtilis strain (EU07) against *Fusarium oxysporum* f. sp. radicis-lycopersici. Physiological and Molecular Plant Pathology. 2008;**73**(1):25-32

[10] Rowe RC, Farley JD, David L. Coplin. Airborne Spore Dispersal and Recolonization of Steamed Soil by *Fusarium oxysporum* in Tomato Greenhouses. 1977;**67**(12):1513-1517

[11] Tomer A, Singh R, Prasad D. Compatibility *Trichoderma harzianum* with systemic and two non-systemic fungicides of *in vitro*. Asian Journal of Crop Science. 2018;**10**(4):174-179

[12] Okigbo RN, Ikediugwu FEO. Studies on biological control of postharvest rot in yams (*Dioscorea* spp.) using *Trichoderma viride*. Phytopathology. 2000;**148**(6):351-355

[13] Lee BH. Advanced Fermentation and Cell Technology. Vol. 2. Hoboken, New Jersey, US: John Wiley & Sons; 2021

[14] Tumbarski Y, Deseva I, Mihaylova D, Stoyanova M, Krastev L, Nikolova R, et al. Isolation, characterization and amino acid composition of a bacteriocin

produced by *Bacillus methylotrophicus* strain BM47. Food Science and Biotechnology. 2018;**56**(4):546

[15] Ravel J, Fraser CM. Genomics at the genus scale. Trends in Microbiology. 2005;**13**(3):95-97

[16] Sultana F, Islam MR, Sabbir MA, Hossain MM. Plant growth promotion and suppression of damping off in tomato by plant growth promoting *Rhizobacterium Bacillus*. Crops. 2020;**5**(1):59-68

[17] Thomloudi EE, Tsalgatidou PC, Baira E, Papadimitriou K, Venieraki A, Katinakis P. Genomic and metabolomic insights into secondary metabolites of the novel bacillus halotolerans Hil4, an endophyte with promising antagonistic activity against gray mold and plant growth promoting potential. Microorganisms. 2021;**9**(12):2508

[18] Ramírez V, Martínez J, Bustillos-Cristales MDR, Catañeda-Antonio D, Munive JA, Baez A. Bacillus cereus MH778713 elicits tomato plant protection against *Fusarium oxysporum*. Journal of Applied Microbiology. 2022;**132**(1):470-482

[19] Emmert EA, Handelsman J. Biocontrol of plant disease: A (Gram-) positive perspective. FEMS Microbiology Letters. 1999;**171**(1):1-9

[20] Ryu CM, Farag MA, Hu CH, Reddy MS, Kloepper JW, Paré PW. Bacterial volatiles induce systemic resistance in Arabidopsis. Plant Physiology. 2004;**134**(3):1017-1026

[21] Lee JP, Lee SW, Kim CS, Son JH, Song JH, Lee KY, et al. Evaluation of formulations of *Bacillus licheniformis* for the biological control of tomato gray mold caused by *Botrytis cinerea*. Biological Control. 2006;**37**(3):329-337

[22] Khan N, Martínez-Hidalgo P, Ice TA, Maymon M, Humm EA, Nejat N, et al. Antifungal activity of *Bacillus* species against *Fusarium* and analysis of the potential mechanisms used in biocontrol. Frontiers in Microbiology. 2018;**9**:2363

[23] Singh D, Devappa V, Yadav DK. Suppression of tomato bacterial wilt incited by *Ralstonia pseudosolanacearum* using polyketide antibiotic-producing Bacillus spp. isolated from rhizospheric soil. Agriculture. 2022;**12**(12):2009

[24] Chou HP, Huang YC, Lin YH, Deng WL. Selection, formulation, and field evaluation of *Bacillus amyloliquefaciens* PMB01 for its application to manage tomato bacterial wilt disease. Agriculture. 2022;**12**(10):1714

[25] Yan Y, Xu W, Hu Y, Tian R, Wang Z. Bacillus velezensis YYC promotes tomato growth and induces resistance against bacterial wilt. Biological Control. 2022;**172**:104977

[26] Phae CG, Shoda M, Kita N, Nakano M, Ushiyama K. Biological control of crown and root rot and bacterial wilt of tomato by *Bacillus subtilis* NB22. Japanese Journal of Phytopathology. 1992;**58**(3):329-339

[27] Im SM, Yu NH, Joen HW, Kim SO, Park HW, Park AR, et al. Biological control of tomato bacterial wilt by oxydifficidin and difficidin-producing *Bacillus methylotrophicus* DR-08. Pesticide Biochemistry and Physiology. 2020;**163**:130-137

[28] Athira S, Anith KN. Plant growth promotion and suppression of bacterial wilt incidence in tomato by rhizobacteria, bacterial endophytes and the root endophytic fungus *Piriformospora indica*. Indian Phytopathology. 2020;**73**(4):629-642

[29] Chen X, Wang Y, Gao Y, Gao T, Zhang D. Inhibitory abilities of Bacillus isolates and their culture filtrates against the gray mold caused by *Botrytis cinerea* on postharvest fruit. Plant Pathology. 2019;**35**(5):425

[30] Zhou L, Song C, Muñoz CY, Kuipers OP. *Bacillus cabrialesii* BH5 protects tomato plants against *Botrytis cinerea* by production of specific antifungal compounds. Frontiers in Microbiology. 2021;**12**:707609

[31] Mari M, Guizzardi M, Brunelli M, Folchi A. Postharvest biological control of grey mould (*Botrytis cinerea* Pers.: Fr.) on fresh-market tomatoes with *Bacillus amyloliquefaciens*. Crop Protection. 1996;**15**:699-705

[32] Yu YY, Si FJ, Wang N, Wang T, Jin Y, Zheng Y, et al. Bacillus-secreted oxalic acid induces tomato resistance against gray mold disease caused by *Botrytis cinerea* by activating the JA/ET pathway. Molecular Plant-Microbe Interactions. 2022;**35**(8):659-671

[33] Ji X, Li J, Meng Z, Zhang S, Dong B, Qiao K. Synergistic effect of combined application of a new fungicide fluopimomide with a biocontrol agent *Bacillus methylotrophicus* TA-1 for management of gray mold in tomato. Plant Disease. 2019;**103**(8):1991-1997

[34] Bu S, Munir S, He P, Li Y, Wu Y, Li X, et al. *Bacillus subtilis* L1-21 as a biocontrol agent for postharvest gray mold of tomato caused by *Botrytis cinerea*. Biological Control. 2021;**157**:104568

[35] Zhou L, Song C, Li Z, Kuipers OP. Antimicrobial activity screening of rhizosphere soil bacteria from tomato and genome-based analysis of their antimicrobial biosynthetic potential. BMC Genomics. 2021;**22**:1-14

[36] Ma Z, Hu J. Plipastatin A1 produced by a marine sediment-derived Bacillus amyloliquefaciens SH-B74 contributes to the control of gray mold disease in tomato. 3 Biotech. 2018;**8**:1-10

[37] Akram W, Anjum T, Ali B. Searching ISR determinant/s from *Bacillus subtilis* IAGS174 against Fusarium wilt of tomato. Biological Control. 2015;**60**:271-280

[38] Devi NO, Tombisana Devi RK, Debbarma M, Hajong M, Thokchom S. Effect of endophytic Bacillus and arbuscular mycorrhiza fungi (AMF) against Fusarium wilt of tomato caused by *Fusarium oxysporum* f. sp. lycopersici. Egyptian Journal of Biological Pest Control. 2022;**32**(1):1-14

[39] Ajilogba CF, Babalola OO, Ahmad F. Antagonistic effects of Bacillus species in biocontrol of tomato Fusarium wilt. Studies on Ethno-medicine. 2013;**7**(3):205-216

[40] Zhang Z, Li J, Zhang Z, Liu Y, Wei Y. Tomato endophytic bacteria composition and mechanism of suppressiveness of wilt disease (*Fusarium oxysporum*). Frontiers in Microbiology. 2021;**12**:731764

[41] Elanchezhiyan K, Keerthana U, Nagendran K, Prabhukarthikeyan SR, Prabakar K, Raguchander T, et al. Multifaceted benefits of *Bacillus amyloliquefaciens* strain FBZ24 in the management of wilt disease in tomato caused by *Fusarium oxysporum* f. sp. *lycopersici*. Physiological and Molecular Plant Pathology. 2018;**103**:92-101

[42] Dhouib H, Zouari I, Abdallah DB, Belbahri L, Taktak W, Triki MA, et al. Potential of a novel endophytic *Bacillus velezensis* in tomato growth promotion and protection against *Verticillium*

wilt disease. Biological Control. 2019;**139**:104092

[43] Akram W, Mahboob A, Javed AA. Bacillus thuringiensis strain 199 can induce systemic resistance in tomato against *Fusarium wilt*. European Journal of Microbiology and Immunology. 2013;**3**(4):275-280

[44] Heidarzadeh N, Baghaee-Ravari S. Application of *Bacillus pumilus* as a potential biocontrol agent of *Fusarium wilt* of tomato. Archives of Phytopathology and Plant Protection. 2015;**48**(13-16):841-849

[45] Alippi AM, Bó ED, Ronco LB, Casanova PE, Aguilar OM. Tomato as a new host of *Erwinia carotovora* subsp. *carotovora* in Argentina. Plant Disease. 1997;**81**(2):230-230

[46] Chandrasekaran M, Chun SC. Expression of PR-protein genes and induction of defense-related enzymes by *Bacillus subtilis* CBR05 in tomato (*Solanum lycopersicum*) plants challenged with *Erwinia carotovora* subsp. *carotovora*. Bioscience, Biotechnology, and Biochemistry. 2016;**80**(11):2277-2283

[47] Lanna-Filho R, Souza RM, Alves E. Induced resistance in tomato plants promoted by two endophytic bacilli against bacterial speck. Tropical Plant Pathology. 2017;**42**:96-108

[48] Buell CR, Joardar V, Lindeberg M, Selengut J, Paulsen IT, Gwinn ML, et al. The complete genome sequence of the Arabidopsis and tomato pathogen *Pseudomonas syringae* pv. tomato DC3000. Proceedings of the National Academy of Sciences. 2003;**100**(18):10181-10186

[49] Ferraz HGM, Resende RS, Moreira PC, Silveira PR, Milagres EA, Oliveira JR, et al. Antagonistic rhizobacteria and jasmonic acid induce resistance against tomato bacterial spot. Bragantia. 2015;**74**:417-427

[50] Roberts PD, Momol MT, Ritchie L, Olson SM, Jones JB, Balogh B. Evaluation of spray programs containing famoxadone plus cymoxanil, acibenzolar-S-methyl, and *Bacillus subtilis* compared to copper sprays for management of bacterial spot on tomato. Crop Protection. 2008;**27**(12):1519-1526

[51] Umesha S, Hariprasad P. Plant health improvement by *Bacillus subtilis* strain GBO3 in tomato against bacterial spot disease. Indian Phytopathology. 2010;**63**(2):127

[52] Yi HS, Yang JW, Ryu CM. ISR meets SAR outside: Additive action of the endophyte Bacillus pumilus INR7 and the chemical inducer, benzothiadiazole, on induced resistance against bacterial spot in field-grown pepper. Frontiers in Plant Science. 2013;**4**:122

[53] Mates ADPK, de Carvalho Pontes N, de Almeida Halfeld-Vieira B. *Bacillus velezensis* GF267 as a multi-site antagonist for the control of tomato bacterial spot. Biological Control. 2019;**137**:104013

[54] Chien YC, Huang CH. Biocontrol of bacterial spot on tomato by foliar spray and growth medium application of *Bacillus amyloliquefaciens* and *Trichoderma asperellum*. European Journal of Plant Pathology. 2020;**156**(4):995-1003

[55] Asaka O, Shoda M. Biocontrol of *Rhizoctonia solani* damping-off of tomato with Bacillus subtilis RB14. Applied and Environmental Microbiology. 1996;**62**(11):4081-4085

[56] Jayaraj J, Radhakrishnan NV, Kannan R, Sakthivel K, Suganya D, Venkatesan S, et al. Development of new formulations of Bacillus subtilis for

management of tomato damping-off caused by *Pythium aphanidermatum*. Biocontrol Science and Technology. 2005;**15**(1):55-65

[57] Jeyaseelan EC, Tharmila S, Niranjan K. Antagonistic activity of *Trichoderma* spp. and Bacillus spp. against Pythium aphanidermatum isolated from tomato damping off. Archives of Applied Science Research. 2012;**4**(4):1623-1627

[58] Zouari I, Jlaiel L, Tounsi S, Trigui M. Biocontrol activity of the endophytic *Bacillus amyloliquefaciens* strain CEIZ-11 against *Pythium aphanidermatum* and purification of its bioactive compounds. Biological Control. 2016;**100**:54-62

[59] Karima HEH, Nadia GE. In vitro study on *Fusarium solani* and *Rhizoctonia solani* isolates causing the damping off and root rot diseases in tomatoes. Nature Physical Science. 2012;**10**(11):16-25

[60] Mezeal IA. Study biocontrol efficacy of *Pseudomonas fluorescens* and *Bacillus subtilis* against *Rhizoctonia solani* and *Fusarium oxysporum* causing disease in tomato (*Lycopersicon esculentum* L.). Indian Journal of Fundamental and Applied Life Sciences. 2014;**4**(4):175-183

[61] Hollensteiner J, Wemheuer F, Harting R, Kolarzyk AM, Diaz Valerio SM, Poehlein A, et al. *Bacillus thuringiensis* and *Bacillus weihenstephanensis* inhibit the growth of phytopathogenic Verticillium species. Frontiers in Microbiology. 2017;**7**:2171

[62] Pei D, Zhang Q, Zhu X, Zhang L. Biological control of *Verticillium wilt* and growth promotion in tomato by rhizospheric soil-derived *Bacillus amyloliquefaciens* Oj-2.16. Pathogens. 2023;**12**(1):37

[63] Bahramisharif A, Rose LE. Efficacy of biological agents and compost on growth and resistance of tomatoes to late blight. Planta. 2019;**249**:799-813

[64] Chowdappa P, Kumar SM, Lakshmi MJ, Upreti KK. Growth stimulation and induction of systemic resistance in tomato against early and late blight by *Bacillus subtilis* OTPB1 or *Trichoderma harzianum* OTPB3. Biological Control. 2013;**65**(1):109-117

[65] Basamma RH, Shripad K. Prevalence of *Bacillus subtilis* in rhizosphere and severity of different diseases of tomato in northern Karnataka. Karnataka Journal of Agricultural Sciences. 2015;**28**(3):351-355

[66] Silva HSA, Romeiro RS, Carrer Filho R, Pereira JLA, Mizubuti ESG, Mounteer A. Induction of systemic resistance by *Bacillus cereus* against tomato foliar diseases under field conditions. Journal of Phytopathology. 2004;**152**(6):371-375

[67] Yan Z, Reddy MS, Ryu CM, McInroy JA, Wilson M, Kloepper JW. Induced systemic protection against tomato late blight elicited by plant growth-promoting rhizobacteria. Phytopathology. 2002;**92**(12):1329-1333

[68] Oancea F, Cornea P, Popa G, Sicuia O, Draganoiu M, Dinu S, et al. *Bacillus amyloliquefaciens* strain antagonistic to *Erwinia amylovora* and elicitor of tomato defense response. IOBC/WPRS Bulletin. 2009;**43**:172

[69] Pane C, Zaccardelli M. Evaluation of Bacillus strains isolated from solanaceous phylloplane for biocontrol of Alternaria early blight of tomato. Biological Control. 2015;**84**:11-18

[70] Modolon TA, Boff P, Boff MIC, Miquelluti DJ. Homeopathic and high dilution preparations for pest management to tomato crop under

organic production system. Horticultura Brasileira. 2012;**30**:51-57

[71] Yamamoto S, Shiraishi S, Kawagoe Y, Mochizuki M, Suzuki S. Impact of *Bacillus amyloliquefaciens* S13-3 on control of bacterial wilt and powdery mildew in tomato. Pest Management Science. 2015;**71**(5):722-727

[72] Ni L, Punja ZK. Management of fungal diseases on cucumber (*Cucumis sativus* L.) and tomato (*Solanum lycopersicum* L.) crops in greenhouses using *Bacillus subtilis*. Bacilli and Agrobiotechnology: Phytostimulation and Biocontrol. 2019;**2**:1-28

[73] Saidi N, Kouki S, M'Hiri F, Hajlaoui MR, Mahrouk M, Ouzari H, et al. Characterization and selection of *Bacillus* sp. strains, effective biocontrol agents against *Fusarium oxysporum* f. sp. radicis-lycopersici, the causal agent of Fusarium crown and root rot in tomato. Annales de Microbiologie. 2009;**59**:191-198

[74] Rashad YM, Abdalla SA, Sleem MM. Endophytic *Bacillus subtilis* SR22 triggers defense responses in tomato against rhizoctonia root rot. Plants. 2022;**11**(15):2051

[75] Samaras A, Efthimiou K, RoumeliotisE,KaraoglanidisGS.Biocontrol potential and plant-growth-promoting effects of *Bacillus amyloliquefaciens* MBI 600 against *Fusarium oxysporum* f. sp. radicis-lycopersici on tomato. In: V International Symposium on Tomato Diseases: Perspectives and Future Directions in Tomato Protection 1207. Leuven, Belgium: International Society for Horticultural Science (ISHS); 2016. pp. 139-146

[76] Sharma R, Chauhan A, Shirkot CK. Characterization of plant growth promoting Bacillus strains and their potential as crop protectants against *Phytophthora capsici* in tomato. Biological Agriculture and Horticulture. 2015;**31**(4):230-244

[77] Beecher GR. Nutrient content of tomatoes and tomato products. Proceedings of the Society for Experimental Biology and Medicine. 1998;**218**(2):98-100

[78] Yoshimura M, Toyoshi T, Sano A, Izumi T, Fujii T, Konishi C, et al. Antihypertensive effect of a γ-aminobutyric acid rich tomato cultivar 'DG03-9'in spontaneously hypertensive rats. Journal of Agricultural and Food Chemistry. 2010;**58**(1):615-619

[79] Yamamoto I, Komada H, Kyniyasu K, Saito M, Ezuka A. A new race of *Fusarium oxysporum* f. sp. lycopersici inducing root rot of tomato. In: Proceedings of the Kansai Plant Protection Society. 1974;**16**:17-29

[80] Manzo D, Ferriello F, Puopolo G, Zoina A, D'Esposito D, Tardella L, et al. *Fusarium oxysporum* f. sp. *radicis-lycopersici* induces distinct transcriptome reprogramming in resistant and susceptible isogenic tomato lines. BMC Plant Biology. 2016;**16**:1-14

[81] Can C, Yucel S, Korolev N, Katan T. First report of fusarium crown and root rot of tomato caused by *Fusarium oxysporum* f. sp. radicis-lycopersici in Turkey. Plant Pathology. 2004;**53**:6

[82] Katan T, Katan J. Vegetative compatibility grouping in *Fusarium oxysporum* f. sp. radicis-lycopersici from the UK, the Netherlands, Belgium and France. Plant Pathology. 1999;**48**(4):541-549

[83] Deng YY, Li W, Zhang P, Sun HY, Zhang XX, Zhang AX, et al. *Fusarium pseudograminearum* as an emerging

pathogen of crown rot of wheat in eastern China. Plant Pathology. 2020;**69**(2):240-248

[84] Lamichhane JR, Dürr C, Schwanck AA, Robin MH, Sarthou JP, Cellier V, et al. Integrated management of damping-off diseases. A review. Agronomy for Sustainable Development. 2017;**37**:1-25

[85] Kemerait RC. P. Vidhyasekaran, Concise encyclopedia of Plant Pathol, Food Products Press, 2006. Binghampton, NY, USA (2004) ISBN 1-56022-943-8 p. 619. Price $79.95 (softback); $129.95 (hardback); ISBN 1-56022-943-8. Agricultural Systems. 2006;**87**(1):120-122

[86] Horst RK, editor. Westcott's Plant Disease Handbook. Berlin, Germany: Springer Science & Business Media; 2013

[87] Sturrock CJ, Woodhall J, Brown M, Walker C, Mooney SJ, Ray RV. Effects of damping-off caused by *Rhizoctonia solani* anastomosis group 2-1 on roots of wheat and oil seed rape quantified using X-ray Computed Tomography and real-time PCR. Frontiers in Plant Science. 2015;**6**:461

[88] Kato M, Minamida K, Tojo M, Kokuryu T, Hamaguchi H, Shimada S. Association of Pythium and Phytophthora with pre-emergence seedling damping-off of soybean grown in a field converted from a paddy field in Japan. Plant Production Science. 2013;**16**(1):95-104

[89] Kumar SP, Srinivasulu A, Babu KR. Symptomology of major fungal diseases on tomato and its management. Journal of Pharmacognosy and Phytochemistry. 2018;**7**(6):1817-1821

[90] Halo BA, Al-Yahyai RA, Maharachchikumbura SS, Al-Sadi AM. *Talaromyces variabilis* interferes with *Pythium aphanidermatum* growth and suppresses Pythium-induced damping-off of cucumbers and tomatoes. Scientific Reports. 2019;**9**(1):1-10

[91] Wiam B, Wassima L, Abderrahmene D, Omar G. Experimental approach for easy identify fusarium wilt of tomato. World. 2019;**8**(2):44-48

[92] Madbouly AK, Abdel-Aziz MS, Abdel-Wahhab MA. Biosynthesis of nanosilver using *Chaetomium globosum* and its application to control Fusarium wilt of tomato in the greenhouse. IET Nanobiotechnology. 2017;**11**(6):702-708

[93] Anil Kumar R, Raj Kumar HG. In vitro antifungal activity of some plant extracts against *Fusarium oxysporum* f. sp. lycopersici. Asian Journal of Plant Sciences. 2015;**5**(1):22-27

[94] Dodds GT. Colour Changes during Chilling of Tomato (*Lycopersicon esculentum* Mill.) Fruit, Surface Topology of Injury and Some Physiological Changes. Ithaca, New York, United States: Cornell University; 1991

[95] Senthamilselvi S, Victoria J. An overview of wilt diseases of tomato. International Journal of Pharmacy and Biological Sciences. 2019;**9**:360-369

[96] Fradin EF, Thomma BP. Physiology and molecular aspects of *Verticillium wilt* diseases caused by *V. dahliae* and *V. albo-atrum*. Molecular Plant Pathology. 2006;**7**(2):71-86

[97] Van der Does HC, Constantin ME, Houterman PM, Takken FLW, Cornelissen BJC, Haring MA, et al. *Fusarium oxysporum* colonizes the stem of resistant tomato plants, the extent varying with the R-gene present. European Journal of. Plant Pathology. 2019;**154**:55-65

[98] Ali M, Cheng Z, Ahmad H, Hayat S. Reactive oxygen species (ROS) as defenses against a broad range of plant fungal infections and case study on ROS employed by crops against *Verticillium dahliae* wilts. Journal of Plant Interactions. 2018;**13**(1):353-363

[99] Sanoubar R, Barbanti L. Fungal diseases on tomato plant under greenhouse condition. European Journal of Biological Research. 2017;**7**(4):299-308

[100] López-Escudero FJ, Mercado-Blanco J. *Verticillium wilt* of olive: A case study to implement an integrated strategy to control a soil-borne pathogen. Plant and Soil. 2011;**344**:1-50

[101] Harvey JM, Smith WL, Kaufman J. Market Diseases of Stone Fruits: Cherries, Peaches, Nectarines, Apricots and Plums (No. 414). Washinton, DC: Agricultural Research Service, US Department of Agriculture; 1972

[102] Williamson B, Tudzynski B, Tudzynski P, Van Kan JA. *Botrytis cinerea*: The cause of grey mould disease. Molecular Plant Pathology. 2007;**8**:561-580

[103] Bojkov G. Gray Mold (*Botrytis Cinerea*) at Vines. USA: Scientific Research Publishing, Inc; 2022

[104] Elad Y, Williamson B, Tudzynski P, Delen N, editors. Botrytis: Biology, Pathology and Control. Berlin, Heidelberg, Germany: Springer Science & Business Media; 2004

[105] Judelson HS. Dynamics and innovations within oomycete genomes: Insights into biology, pathology, and evolution. Eukaryotic Cell. 2012;**11**(11):1304-1312

[106] Fry WE, Birch PRJ, Judelson HS, Grünwald NJ, Danies G, Everts KL, et al. Five reasons to consider *Phytophthora infestans* a reemerging pathogen. Phytopathology. 2015;**105**(7):966-981

[107] Nelson SC. Late Blight of Tomato (*Phytophthora infestans*). 2008;**45**:1-10

[108] Saville A, Graham K, Grünwald NJ, Myers K, Fry WE, Ristaino JB. Fungicide sensitivity of US genotypes of *Phytophthora infestans* to six oomycete-targeted compounds. Plant Disease. 2015;**99**(5):659-666

[109] Arakeri MP, Arun M, Padmini RK. Analysis of late blight disease in tomato leaf using image processing techniques. International Journal of Precision Engineering and Manufacturing. 2015;**5**(4):12-22

[110] Solankey SS, Ray PK, Kumari M, Singh HK, Shamim M, Verma DK, et al. Tomato Diseases, Their Impact, and Management. In: Biotic Stress Management in Tomato. New Jersey, Canada: Apple Academic Press; 2021. pp. 1-23

[111] Haverkort AJ, Struik PC, Visser RGF, Jacobsen EJPR. Applied biotechnology to combat late blight in potato caused by *Phytophthora infestans*. Potato Research. 2009;**52**:249-264

[112] Bessadat N, Berruyer R, Hamon B, Bataille-Simoneau N, Benichou S, Kihal M, et al. Alternaria species associated with early blight epidemics on tomato and other Solanaceae crops in northwestern Algeria. European Journal of Plant Pathology. 2017;**148**:181-197

[113] Sharma RK, Patel DR, Chaudhari DR, Kumar V, Patel MM. Effect of some fungicides against early blight of tomato (*Lycopersicon esculentum* Mill.) caused by *Alternaria solani* (Ell. & Mart.) Jones and Grout and their impact on yield. International Journal of Current

Microbiology and Applied Sciences. 2018;7(7):1395-1401

[114] Van der Waals JE, Korsten L, Aveling TAS. A review of early blight of potato. African Plant Protection. 2001;7(2):91-102

[115] Sreenivasulu R, Reddy MSP, Tomar DS, Sanjay MSS, Reddy BB. Managing of early blight of tomato caused by *Alternaria solani* through fungicides and bioagents. International Journal of Current Microbiology and Applied Sciences. 2019;**8**(6):1442-1452

[116] Hassan EA, Hifney AF, Ali EF, Sayed AM. Fungicidal activities and molecular docking of the marine alga *Ulva lactuca* and *Punica granatum* peel extracts on Alternaria tomato spot disease. Environmental Science and Pollution Research. 2023;**30**(8):21872-21887

[117] Delahaut KA, Newenhouse AC. Growing Tomatoes, Peppers, and Eggplants in Wisconsin. Madison, Wisconsin, USA: University of Wisconsin Ext; 1997

[118] Wale S, Platt B, Cattlin ND. Diseases, Pests and Disorders of Potatoes: A colour Handbook. Boca Raton, Florida, United States: CRC Press; 2008

[119] Jacob D, David DR, Sztjenberg A, Elad Y. Conditions for development of powdery mildew of tomato caused by *Oidium neolycopersici*. Phytopathology. 2008;**98**(3):270-281

[120] Whipps JM, Budge SP, Fenlon JS. Characteristics and host range of tomato powdery mildew. Plant Pathology. 1998;**47**(1):36-48

[121] Jones H, Whipps JM, Gurr SJ. The tomato powdery mildew fungus *Oidium neolycopersici*. Molecular Plant Pathology. 2001;**2**(6):303-309

[122] Glawe DA. The powdery mildews: A review of the world's most familiar (yet poorly known) plant pathogens. Annual Review of Phytopathology. 2008;**46**:27-51

[123] Levente KISS, Roger TA, Saenz GS, Cunnington JH, Takamatsu S, Pascoe I, et al. Identification of two powdery mildew fungi, *Oidium neolycopersici* sp. nov. and *O. lycopersici*, infecting tomato in different parts of the world. Mycological Research. 2001;**105**(6):684-697

[124] Matsuda Y, Kashimoto K, Takikawa Y, Aikami R, Nonomura T, Toyoda H. Occurrence of new powdery mildew on greenhouse tomato cultivars. Journal of General Plant Pathology. 2001;**67**:294-298

[125] Gleason ML, Edmunds BA. Tomato Diseases and Disorders. Ames, IA: Iowa State University, University Extension; 2005

[126] Bharate AA, Shirdhonkar MS. A review on plant disease detection using image processing. In: 2017 International Conference on Intelligent Sustainable Systems (ICISS). New York, United States: IEEE; 2017. pp. 103-109

[127] Sun G, Feng C, Guo J, Zhang A, Xu Y, Wang Y, et al. The tomato Arp2/3 complex is required for resistance to the powdery mildew fungus *Oidium neolycopersici*. Plant, Cell & Environment. 2019;**42**(9):2664-2680

[128] Selby AD. Investigations of Plant Diseases in Forcing House and Garden. 1986;**73**:221-226

[129] Ávila MCR, Lourenco V Jr, Quezado-Duval AM, Becker WF, de Abreu-Tarazi MF, Borges LC, et al. Field

validation of TOMCAST modified to manage Septoria leaf spot on tomato in the central-west region of Brazil. Crop Protection. 2020;**138**:105333

[130] Cabral RN, Marouelli WA, Lage DA, Café-Filho AC. Septoria leaf spot in organic tomatoes under diverse irrigation systems and water management strategies. Horticultura Brasileira. 2013;**31**:392-400

[131] Reis A, Lopes C. Oídios do tomateiro. Brasília, Brazil: Embrapa Hortaliças; 2009

[132] Gul Z, Ahmed M, Khan ZU, Khan B, Iqbal M. Evaluation of tomato lines against Septoria leaf spot under field conditions and its effect on fruit yield. Agricultural Sciences. 2016;**7**(4):181-186

[133] Nayyer MA, Ahmad MF, Shamim M, Lal D, Srivastava D, Tripathi VK. Molecular approaches for the control of Septoria leaf spot in tomato. In: Biotic Stress Management in Tomato. New Jersey, Canada: Apple Academic Press; 2021. pp. 91-106

[134] Mani SD, Pandey S, Govindan M, Muthamilarasan M, Nagarathnam R. Transcriptome dynamics underlying elicitor-induced defense responses against Septoria leaf spot disease of tomato (*Solanum lycopersicum* L.). Physiology and Molecular Biology of Plants. 2021;**27**:873-888

[135] Mattos ADP, Tolentino Júnior JB, Itako AT. Determination of the severity of Septoria leaf spot in tomato by using digital images. Australasian Plant Pathology. 2020;**49**:329-356

[136] Thomma BP, Van Esse HP, Crous PW, De Wit PJ. *Cladosporium fulvum* (syn. *Passalora fulva*), a highly specialized plant pathogen as a model for functional studies on plant pathogenic Mycosphaerellaceae. Molecular Plant Pathology. 2005;**6**(4):379-393

[137] Cooke MC. New American fungi. Grevillea. 1883;**12**:22-33

[138] Li S, Zhao T, Li H, Xu X, Li J. First report of races 2.5 and 2.4. 5 of *Cladosporium fulvum* (syn. *Passalora fulva*), causal fungus of tomato leaf mold disease in China. Journal of General Plant Pathology. 2015;**81**:162-165

[139] Lucentini CG, Medina R, Franco ME, Saparrat MC, Balatti PA. *Fulvia fulva* [syn. *Cladosporium fulvum*, *Passalora fulva*] races in Argentina are evolving through genetic changes and carry polymorphic avr and ecp gene sequences. European Journal of Plant Pathology. 2021;**159**(3):525-542

[140] De Wit PJ, Van Der Burgt A, Ökmen B, Stergiopoulos I, Abd-Elsalam KA, Aerts AL, et al. The genomes of the fungal plant pathogens *Cladosporium fulvum* and *Dothistroma septosporum* reveal adaptation to different hosts and lifestyles but also signatures of common ancestry. PLoS Genetics. 2012;**8**(11):e1003088

[141] Akbar A, Din S, Ahmad M, Khan G, Alam S. Effect of Phytobiocides in controlling soft rot of tomato. Journal of Natural Science Research. 2014;**4**(11):2225-09921

[142] Bell KS, Sebaihia M, Pritchard L, Holden MTG, Hyman LJ, Holeva MC, et al. Genome sequence of the enterobacterial phytopathogen *Erwinia carotovora* subsp. *atroseptica* and characterization of virulence factors. Proceedings of the National Academy of Sciences. 2004;**101**(30):11105-11110

[143] Malathrakis NE, Goumas DE. Bacterial soft rot of tomato in plastic

greenhouses in Crete. The Annals of Applied Biology. 1987;**111**(1):115-123

[144] Akbar A, Ahmad M, Khan SZ, Ahmad Z. Characterization of the causal organism of soft rot of tomatoes and other vegetables and evaluation of its most aggressive isolates. American Journal of Plant Sciences. 2015;**6**(4):511

[145] Adamu SH, Lal AA, Simon S. In vitro efficacy of certain botanicals against bacterial soft rot of tomato (*Solanum lycopersicum* L.). African Journal of Agricultural Research. 2017;**12**(23):2049-2055

[146] Yunis H, Bashan Y, Okon Y, Henis Y. Weather dependence, yield losses, and control of bacterial speck of tomato caused by Pseudomonas tomato. Plant Disease. 1980;**64**:937-939

[147] Walker JC. Vegetable seed treatment. Botanical Review. 1948;**14**(9):588-601

[148] Louws FJ, Wilson M, Campbell HL, Cuppels DA, Jones JB, Shoemaker PB, et al. Field control of bacterial spot and bacterial speck of tomato using a plant activator. Plant Disease. 2001;**85**(5):481-488

[149] Shenge KC, Mabagala RB, Mortensen CN, Stephan D, Wydra K. First report of bacterial speck of tomato caused by *Pseudomonas syringae* pv. tomato in Tanzania. Plant Disease. 2007;**91**(4):462-462

[150] Da Silva EO, Martins SJ, Alves E. Essential oils for the control of bacterial speck in tomato crop. African Journal of Agricultural Research. 2014;**9**(34):2624-2629

[151] Lukyanenko AN. Disease resistance in tomato. Genetic Improvement of Tomato. 1991;**14**:99-119

[152] Boland GJ, Melzer MS, Hopkin A, Higgins V, Nassuth A. Climate change and plant diseases in Ontario. Canadian Journal of Plant Pathology. 2004;**26**(3):335-350

[153] Horvath DM, Stall RE, Jones JB, Pauly MH, Vallad GE, Dahlbeck D, et al. Transgenic resistance confers effective field level control of bacterial spot disease in tomato. PLoS ONE. 2012;**7**(8):e42036. DOI: 10.1371/journal.pone.0042036

[154] Ortigosa A, Gimenez-Ibanez S, Leonhardt N, Solano R. Design of a bacterial speck resistant tomato by CRISPR/Cas9-mediated editing of Sl JAZ 2. Plant Biotechnology Journal. 2019;**17**(3):665-673

[155] Yabuuchi E, Kosako Y, Yano I, Hotta H, Nishiuchi Y. Transfer of two Burkholderia and an Alcaligenes species to *Ralstonia* gen. nov.: Proposal of *Ralstonia pickettii* (Ralston, Palleroni and Doudoroff 1973) comb. nov., *Ralstonia solanacearum* (Smith 1896) comb. nov. and *Ralstonia eutropha* (Davis 1969) comb. nov. Microbiology and Immunology. 1995;**39**(11):897-904

[156] Murthy KN, Soumya K, Udayashankar AC, Srinivas C, Jogaiah S. Biocontrol potential of plant growth-promoting rhizobacteria (PGPR) against *Ralstonia solanacearum*: Current and future prospects. In: Biocontrol Agents and Secondary Metabolites. Woodhead Publishing; 2021. pp. 153-180

[157] Jones JB, Jones JP, Stall RE, Zitter TA, Hopkins DL, Shew HD. Compendium of Tomato Diseases and Pests. 2nd ed. St. Paul, MN: APS Press; 2014

[158] Cuppels DA, Ainsworth T, Ruggi A. Field crop residue and other potential inoculum sources for the bacterial spot

pathogen in Ontario. Phytopathology. 2008;**98**(6):S43-S44

[159] Abbasi PA, Weselowski B. Efficacy of *Bacillus subtilis* QST 713 formulations, copper hydroxide, and their tank mixes on bacterial spot of tomato. Crop Protection. 2015;**74**:70-76

[160] Jones JB. Bacterial speck. In: Compendium of Tomato Diseases. St. Paul, Minnesota: APS Press; 1991

[161] Tudor SM. Molecular Characterization of Bacteriocin-like Activity in Tomato Race-Three Strains of *Xanthomonas campestris* pv. *vesicatoria*. Gainesville, Florida, USA: University of Florida; 1999

[162] Pawar BT. Report of bacterial diseases of tomato from Marathwada region of Maharashtra, India. Research Journal of Recent Sciences. 2014;**2277**:2502

[163] Venette JR, Lamey HA, Smith RC. Bacterial Spot and Bacterial Speck of Tomato. Fargo, North Dakota, United States: North Dakota State University Extension Services; 1996

[164] Balogh B, Jones JB, Momol MT, Olson SM, Obradovic A, King P, et al. Improved efficacy of newly formulated bacteriophages for management of bacterial spot on tomato. Plant Disease. 2003;**87**(8):949-954

[165] Wang G, Govinden R, Chenia HY, Ma Y, Guo D, Ren G. Suppression of Phytophthora blight of pepper by biochar amendment is associated with improved soil bacterial properties. Biology and Fertility of Soils. 2019;**55**:813-824

[166] Singh M, Patel SK, Kalia VC. Bacillus subtilis as potential producer for polyhydroxyalkanoates. Microbial Cell Factories. 2009;**8**(1):1-11

[167] Lamsal K, Kim SW, Kim YS, Lee YS. Biocontrol of late blight and plant growth promotion in tomato using rhizobacterial isolates. Journal of Microbiology and Biotechnology. 2013;**23**(7):897-904

Chapter 2

Impacting of Root-Knot Nematodes on Tomato: Current Status and Potential Horizons for Its Managing

Mohamed Youssef Banora

Abstract

Root-Knot Nematodes (*Meloidogyne* spp.) are very serious pathogen on tomato plants among the worldwide. They are widely distributed in soil and causes a highly economical losses for more than 5000 plant species. Therefore, many managements' strategies are applicable to decrease their effectiveness such as resistant genotypes, soil solarisation and chemical control. Until now, chemical control is the most applied strategy for nematode management. Although nematicides are highly impacted for nematode suppression but environmentally not safety and very toxic. Consequently, several promising studies revealed that root-knot nematode (RKN) can inhibit nematode reproduction based on the susceptibility of their plant host. The plant effectors play a vital role during nematode infection and effect on plant response to nematode requirements. To understand well the relationship between nematode and their host, the molecular and immunolocalization methods illustrated some proteins which are expressed by plant genes involved in plant–nematode interaction. This chapter will focus on the latest status and future perspectives for nematode management.

Keywords: root-knot nematodes (*Meloidogyne* spp.), tomato (*Solanum lycopersicum*), molecular response, traditional management practices, new approaches for nematode management

1. Introduction

Phyto-parasitic nematode; *Meloidogyne* spp. which are known root-knot nematodes (RKNs) and the most widespread soil-borne obligate plant parasites that are sedentary endoparasites in plant roots [1]. More than identified 100 species belonged under *Meloidogyne* spp. capable infect almost all vascular plant species and distributed in the tropical and subtropical regions [2]. The most common species in the tropical regions are *Meloidogyne incognita*, *M. javanica* and *M. arenaria* while *M. hapla*, *M. fallax* and *M. chitwoodi* are distributed in the cooler regions [3]. Since the humid climate is the most favorable conditions for survival and reproduction of RKNs, the most damage and crop losses pronounced in tropical regions [4]. Tomato plants are grown round the worldwide over all the year and are considered one of the most important hosts for RKNs. Accordingly, these nematodes cause billions of dollars in losses of the tomato crop annually. The success of parasitism and life cycle of these nematodes depends on their induction for nematode-feeding sites (NFS) within the root tissues of plant host [5]. Therefore, root-knot nematode infection formed many tumeres on infected roots called galls that contain NFS within it. Many applicable processes can reduce and manage RNK in filed. Agriculture practices, physical, and biological methods together significantly more effective than pesticides. Unfortunately, nematicides application still used until now since it is the most management strategy to reduce the damage of RKNs. Although, the nematicides highly effect on nematode and infection parameters but the apprehensions on environmental safety and human health risks have led to restriction of the nematicides application, which is commonly expensive and toxic, particularly in sustenance agriculture system. Another approach to inhibit RKNs and their population in soil is the cultivation of resistant plants that depends on natural resistance inherited by resistance genes. However, this management process requires long-term experiments and to date, there are relatively few resistance genes identified. Currently, a new promising approach for suppress RKNs depends on some information became available for about genes that are involved in the relationship between RKNs and their host during infection. Thus, a good understanding of the interaction between plant and nematode enables us to develop a new strategy to reduce the risk of RNKs.

2. Historical brief

Thru the nineteenth century in 1855, Miles Josef Berkeley was the first to record and attributed galls on cucumber roots to nematodes. Later in 1872, Greef titled the pathogen of root galls, as *Heterodera radicicola.* In 1879, Cornu also detected root galls on sainfoin plants (*Onobrychis sativus* Lam.) in the Loire valley, France and described it as root-knot caused by nematode and named it as *Anguillula marioni*. While, in 1884, Müller turn this name to *Heterodera* again. Afterward, Treub in Java, Indonesia named the root gall producing nematode as *Heterodera javanica* in 1885. During 1887, Göeldi described and illustrated a RKN from coffee plants in Brazil briefly and named it *Meloidogyne exigua*. In the meantime, Neal called it as *Anguillula arenaria* in 1889 in the United States of America, while Cobb in 1890 named it as *Tylenchus arenaria* in New Zealand. At the beginning of twentieth century in 1901, Prayer published the first research article relating to RKN that was describing a nematode disease as root galling on banana in Egypt. Then, Kofoid and White have named the root gall inducing nematode as *Oxyuris incognita* in 1919. Generally, the name *Heterodera marioni* was widely used for RKN until 1949, when Chitwood re-established the genus *Meloidogyne* suggested by Göeldi in 1887, and retained four species *M. javanica*, *M. arenaria*, *M. exigua* and *M. incognita* and described *M. hapla* and a variety of *M. incognita* he termed *M. incognita* var. *acrita* [1].

3. Economic impact

Economically, RKN cause several billion dollars of losses annually that is estimated to be totally between US$80 to US$110 billion per year for agriculture crops around the world [6, 7]. The severity of damage and losses caused by *Meloidogyne* depending on the nematode specie, susceptibility of host, crop rotation, season, and soil type [8]. Beside the directly effect of RKN as a plant pathogen, *Meloidogyne* species also able to interact with other soil-borne pathogens especially vascular wilt pathogens

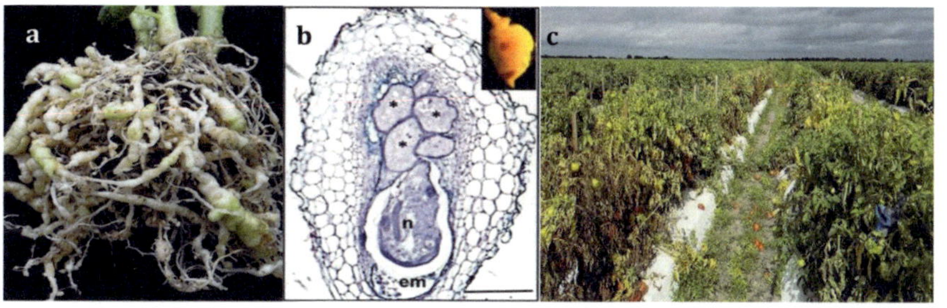

Figure 1.
Symptoms of root-knot nematodes. Galled root system of infected tomato plants (a), histological symptoms of root galls illustrated the developed nematode-feeding site containing adult female (b). Dead, stunted, chlorotic, and preharvest wilting plants caused by root-knot nematode in the field of tomato (c). (*) giant cells, (n) nematode and em = egg-mass. Bars = 200 μm (C) (a, reprinted from DOI: 10.1007/s41348-022-00642-3), (b, reprinted from DOI: 10.24425/jppr.2019.126040) and (c, reprinted from https://www.growingproduce.com/vegetables/field-scouting-guide-root-knot-nematode/#slide=148494-148490-2).

and root-rot pathogens. Consequently, the synergistic effect of RKN and other plant pathogens increase plant damages and crop losses as well. According to the mechanical force of RKN that cause wounds, and their physiological effect on the plant, this interactions effects on the susceptibility and response of host plants to root-rot infection, and lead to breakdown plant resistance particularly for wilt diseases [9, 10]. Tomato (*Solanum lycopersicum* L.) is one of the most popular vegetable crops worldwide and grown on more than 5 million hectares. Annually, tomato plants producing nearly 243 million tons of tomato fruits around the worldwide estimated about US$ 1.6 billion [11]. Commonly, RKN caused more than 85% of the damage to agricultural crops [12], and 68% of tomato yield lost per year [13, 14]. The damage thresholds of *Meloidogyne* depends on their specie, race, and plant type. The average of thresholds has been determined for several crops is approximately 0.5–2 J2/g of soil [15]. Regarding to symptoms caused by RKN infection on their host, many galls formed on plant root system (**Figure 1a**) and counting the nematode feeding sites in the vascular tissues that shelter adult females (**Figure 1b**). Thus, these galls effect on the uptake of nutrients and water by the plant [16]. In addition, plants revealed foliar symptoms such as preharvest wilting, yellowing of leaves, general reduction of plant growth, floral abortions and decrease of both fruits number and quality, as well as death of the plant in severe infections (**Figure 1c**) [17]. Besides the direct losses of *Meloidogyne* spp., the global cost of nematicides marketing is annually developing. The world's increasing focus on controlling the incidences of plant-parasitic nematodes and improving crop yields to ensure food security led to spend US$1.8 billion in 2022. According to increasing the environmental concerns worldwide and improved crop yields, the governments funding is focusing for increase the integrated pest management strategies and to produce new chemical and bio-nematicides as alternatives to traditional synthetic pesticides. Therefore, the cost of creating a new chemical active ingredient is increasing every year and is now estimated to be more than US$250 million [18].

4. Biological life cycle

Stereotypically, nematode's life cycle including six stages; an egg, four juvenile stages and the adult stage. A molting phase occurs between each juvenile and adult stage. Concerning RKN, the parasitic cycle (**Figure 2**) commences when the J2 penetrates a root in the zone of elongation (**Figure 2a**) [19]. Afterwards, J2 successes to move intercellularly through the cortex toward the root tip without causing damage to the root cells (**Figure 2b**). The reason of simplicity penetration and roaming within the root is due to the mechanical force of nematode's stylet and their secretions that including cell-wall-degrading enzymes produced from specialized glands [20–22]. After that, J2 turns around and moves back up into the differentiating vascular cylinder until it reaches the region where the protoxylem is just beginning to form (**Figure 2c**), where it establishes a long-term feeding site. The J2 induces the redifferentiation of five to seven parenchyma root cells for the development of the nematode feeding site structure (**Figure 2d**) [23]. These feeding cells form to multinucleate giant cells inducted by the injection of secretions produced from the dorsal esophageal gland of J2 [24]. When J2 starting feeding, becomes sedentary and directly exchange its shape from a vermiform to fusiform shape after the second and the third molts that differentiates non-feeding phases (J3 and J4) [25]. Then, J4 undergoes the fourth molt to differentiates to adult stage. In optimal conditions, almost all J4 differentiates to young females that feeding resumes again through giant cells. Consequently, the developing females becomes mature and swollen, pear-shaped, and lays approximately 500–2000 eggs embedded and clustered in a gelatinous matrix called egg-mass attached on the root surface (**Figure 2e**). Within the egg (**Figure 2f**), the first stage juvenile (J1) forming

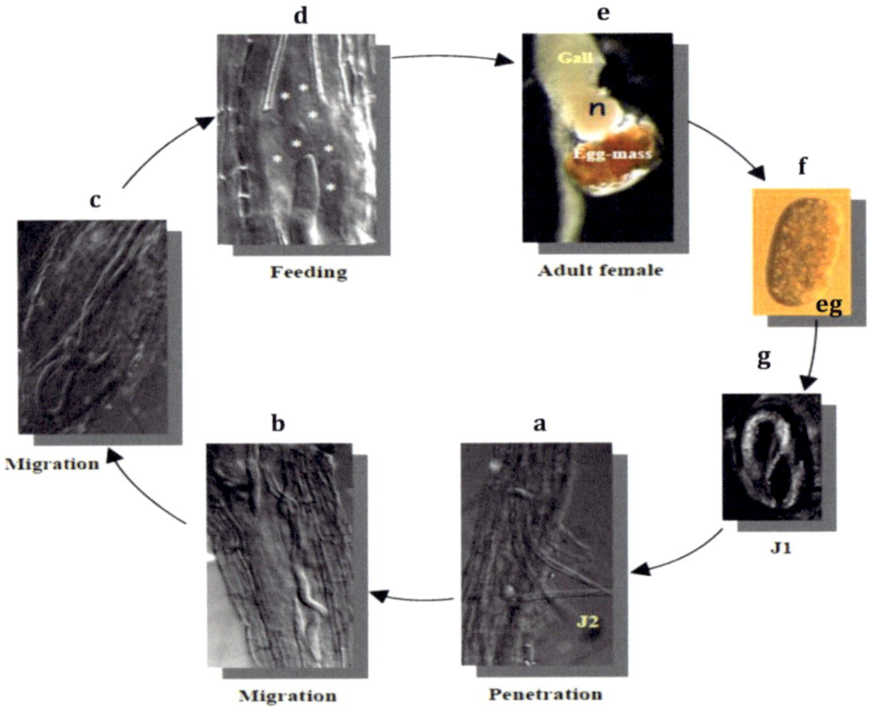

Figure 2.
Root-knot nematode parasitic cycle. The J2 is the only stage that can penetrate the roots closely behind the root tip (a). Intercellularly J2 migrate toward the root tip (b). Behind the root cap, the infective stage (J2) turns-up to the vascular tissue (c). The J2 stops moving and induce approximately 7 feeding cells by its secretions to form nematode feeding site (d) which are rapidly differentiated into multinucleated giant cells (*). Adult female still partially inside the gall and lays its eggs in a gelatinous matrix (egg-mass) (e) attached with root surface. Inside the egg (f) the nematode embryo rapidly develops into J1 (first stage juvenile) (g) that hatch from egg to looking for another plant root to penetrate again.

(**Figure 2g**) and molting to differentiate to second-stage juvenile (J2) which hatches in soil and looking for another plant roots to repeat the disease cycle. Normally this cycle between initial infection and laying egg-masses takes 21 days at 25°C. According to the environmental conditions, plant response and nutrient availability, sometimes during droughty condition or in resistant host, males differentiate and directly leaving the root without feeding [26]. Typically, RKN reproduction by parthenogenesis, although males are frequently found and seem to have no role in sexual reproduction [1].

5. The parasitic approach

According to understanding the plant-nematode interaction relationship, root-knot nematodes have evolved strategies to suppress host immune responses for the development of its feeding sites. The activation of plant immune responses depends on specific molecules that recognize nematode signaling. Recently, genetic sequencing analyses led to identification of molecular components that secreted from RKN during parasitism. These analyses have contributed to our overall understanding of the dynamic and complex nature of plant-nematode interactions. These molecules called efforts and produced in three esophageal salivary glands classified to two subventral glands (SvG) and one dorsal gland (DG). The effectors secreted by SvG allowing J2 penetration and migration in the root while proteins secreted during parasitism are produced by SvG and particularly by DG [27]. Also, some effectors produced in other secretory organs, such as chemosensory amphids [22]. Proteomic analysis has identified around 500 proteins secreted by preparasitic J2s or feeding females of *M. incognita* [28]. Furthermore, during this molecular dialog there is other secreted proteinaceous effectors such as phytohormones, have been shown to favor the plant-nematode interaction [29]. Among proteins secreted by preparasitic J2s, cell wall-degrading effectors have been detected to support its penetration and migration within the root (**Figure 2a**, **b**), and effectors suppressing plant defenses have been described [30, 31]. Root-knot nematodes are sedentary obligate biotrophic pathogens establish a relationship with their host plants, inducing the redifferentiation of root cells into specialized feeding cells for a long-term. Vitally, the successful establishment of nematode feeding cells is critical for nematode development and its reproduction. Nematode feeding cells (NFC) called giant cells (GCs) forming when J2 settle down to start feeding and stimulated by other secreted effectors from J2 that injected via its syringe-like stylet at the beginning of feeding process. Fully differentiated GCs are enlarged cells that estimated around more than 300 times larger than normal cells and are converted into multinucleate cells through synchronous nuclear divisions without cell division (**Figure 3a**) [23]. Therefore, GCs may contain more than a hundred polyploid nuclei that may have undergone extensive endoreduplication (**Figure 3b**) [16, 32]. All these dramatical changes lead to the formation of nematode feeding sites that containing GCs surrounded by active mitotic cells called neighboring cells which lead to display a typical root gall [32–34]. Moreover, the plant cytoskeleton in GCs is totally random as a response to RKN induction during nematode feeding. Both actin filaments and microtubules have observed during RKN infection in *Arabidopsis thaliana* and shown a dense network of cortical microtubules and thick microtubule bundles were formed at giant cell cortex (**Figure 4**), also shown a dense actin network and actin cables illustrated in GC and neighboring cells (**Figure 5**) [35, 36]. Genetically, the formation of GC requires extensive changes to gene expression for several gene involved in the plant-nematode interaction [37].

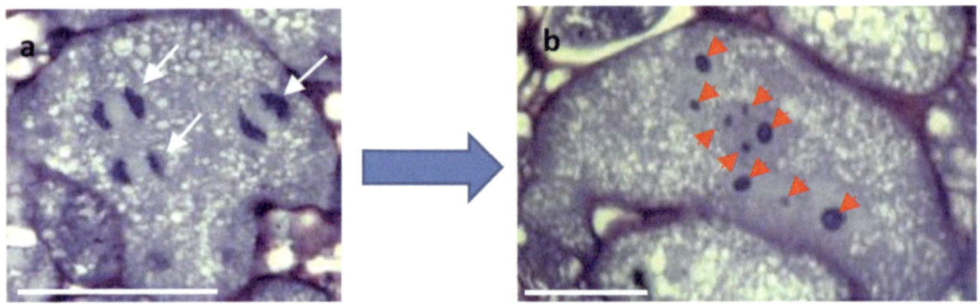

Figure 3.
Giant cell formation. The white arrows indicate to mitotic activity illustrated the nuclear divisions without cytokinesis (a), red arrows indicate to multinuclear in the fully forming giant cells. Bars = 5 µm. DOI: 10.1371/journal.ppat.1002343.

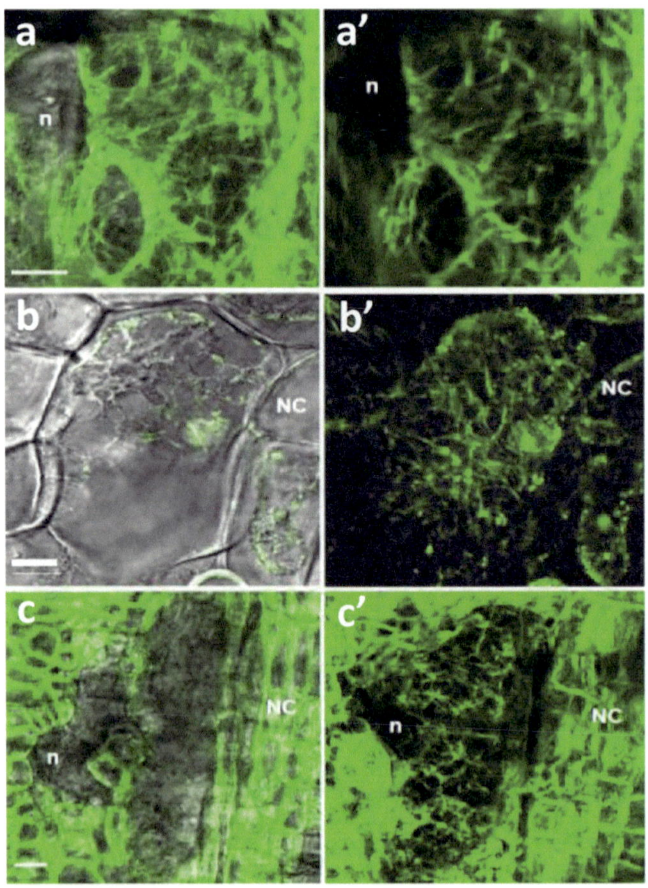

Figure 4.
In vivo observation of cortical microtubules (CMTs) cytoskeleton arrays during giant cell development in roots of Arabidopsis thaliana (microtubule-binding domain MBD-GFP). (A), (B) and (C) overlays of differential interference contrast transmission. (a’), (B′) and (C′) images of confocal laser scanning microscopy showing that a dense network of CMTs and thick microtubule bundles were presented at giant cell cortex. n, nematode; NC, neighboring cell. bars = 5 µm (a - B′), 10 µm (C and C′). DOI: 10.1371/journal.ppat.1002343.

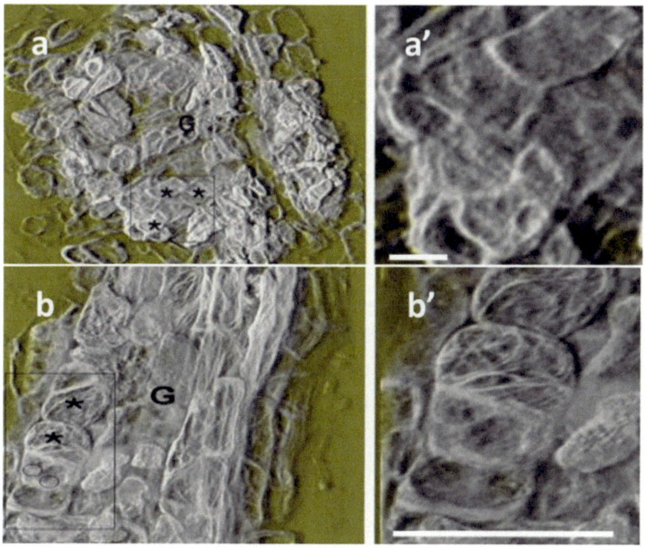

Figure 5.
*Actin cytoskeleton organization in galls inducted by RKN on the root of in roots of Arabidopsis thaliana. Dense actin network in giant cells and neighboring cells forming 10 days after infection (a, a') and 21 days after infection (b, b'). G, gall and *, giant cell. the actin cytoskeleton is depicted in gray levels on a yellow background. Bars = 50 mm. DOI:10.1105/tpc.109.069104.*

Currently, some studies focused on several genes involved in plant response for RKN infection and the development of their NFS that can be promise to a novel possibility horizon for management of *Meloidogyne* spp. in the future.

6. Management strategies

Many applicable processes are impacting on RKN and their damage on plant hosts. Many years ago, until now, the traditional methods including almost all practise related with agronomic methods and chemical control procedures using nematicides are the most familiar techniques for nematode control. Various efforts have focused on this problem in worldwide particularly in developing countries. One of the most widely project for control of *Meloidogyne* spp. was The International *Meloidogyne* Project (IMP) coordinated by North Carolina State University has been performed extensive research for several years to help the developing nations to reduce crop loss caused by RKN. This project has been lunched in 1975 with a teamwork was including more than 100 nematologists participated from more than 70 developing countries. The classical methods for nematode control involve the use of agricultural practices, chemical control using nematicides and resistant plants. Significantly, nematicides are successfully reduce nematode's populations in soil and increase both quantity and quality of agricultural crops. However, they are unsafe highly toxic chemicals consequently, the employed nematicides are highly pollutants and dangerous to our environment. Hence, many alternative chemicals which are eco-friendly can use to increase the immunity of plant host for example some plant extracts. In addition, biological control by using numerous bioagents against nematode can inhibit nematode activity and their production. Commercially, there is some bio-nematicides are common applicable as a substitute for nematicides.

6.1 Traditional practices

6.1.1 Chemical practices

The safety of agroecosystem is facing a major challenge that require protect it from the toxicity of pesticides and increase the outcomes of crops simultaneously. Commonly, chemical controls of RKN using a various nematicides is the most impacting for all *Meloidogyne* spp. and positively decrease crop losses; however, numerous nematicides are being phased out due to environmental and health concerns [7]. Therefore, the alternative chemicals that have both the nematocidal effect and to be eco-friendly is highly required. Recently, several botanical-based nematocidal are being commercially marketed. The extracts of Neem plant (*Azadirachta indica*) have the most famous nematicidal formulations such as Neemrich, Neemix, Neemazal, Neemgold, and Neemax [38]. Also, allicin (diallyl thiosulfinate) extracted from garlic plant (*Allium sativum*) is a nematicidal compound that highly effective against *M. incognita* and improve tomato yield [39]. In parallel, various plant extracts have been synthesized to ensure more safety for ecosystem than the commonly known synthetic chemical nematicides.

6.1.2 Agronomic practices

Agricultural practices are non-chemical management tactics such as crop rotation with non-host crops or resistant cultivars, and these policies are an economical method for nematode management. Although RKN distributed on more than 5000 plant species [40], there are some plants have shown to be poor host such as wheat, corn, sorghum and garlic that can cultivate during crop rotation [7, 41, 42]. The cultivation of non-host crops or resistant cultivars during crop rotation able to suppress RKN populations by decreasing their eggs and infective juveniles (J2) in soil [43]. Concerning tomato crops, the rotation to non-hosts should be for a minimum of 3 years [44]. Also, the exclusion of weed plants is important avoidance strategy for other alternative hosts to reduce nematode population in soil because many weed species may serve as hosts to RKN [45, 46]. Irrigation water is an important facility for nematode transfer. Thus, sanitation of farm equipment, and plant seedling can avoid transferring the pathogen to non-infested fields [47]. Additional procedures such as fallowing soil, soil solarization, steaming, and flooding can decrease the survival rate of nematode's eggs and infective stage (J2) [44]. Furthermore, physical techniques such as soil solarization before planting can be combined with cultural processes for effective control of RKN. The application of soil solarization technique usually during summer season for 8 weeks using transparent polyethylene sheets for wet soil mulching [48]. Under the influence of the sun's heat, water evaporates from the soil to condense on the inner surface of polyethylene sheet in the form of droplets as a lens that collect the sun's rays, which leads to a rise in the soil temperature in the upper 40 cm and reducing both eggs and juveniles survivals [49]. Additional traditional method among agricultural practice is soil amendments with organic manure for improving soil structure, physical and chemical soil properties, temperature, and humidity conditions as well as the nutrient content necessary for plant growth and their immunity for pathogens. The previous studies shown that soil amendment using farm manure and extracts from marigold (*Tagetes* spp.) let to release toxic compounds that can harm plant parasitic nematodes [50] and have also been activated the bio-control agents in soil [47]. The high rate of soil amendments using organic materials has a significant effect on nematode populations in soil [51]. Normally, the indirectly outcomes of soil amendments using organic manures are increasing the activity of many benefit bioagents in soil that can suppress the population levels of many plant pathogens including RKN and may be able to induce systemic resistance of plant species as well [52].

6.1.3 Biological control agents

Biological control methods using living soil-habitat microorganisms (bacteria and fungi) that effect on the nematode's population unites in soil (eggs and/or J2) by secreted the natural bioactive substances [53]. Many studies and experiments focused mainly on bacteria and fungi that are revealed antagonistic effect against RKN. The results of these research achieved to produce some commercial biological products against certain *Meloidogyne* spp. These products are usually developed from bioagents which can attach with nematode cuticle or to parasitize eggs-masses subsequently decreasing nematodes population in soil. Some bacterial isolates shown a highly activity against RKNs infected tomato such as *Pseudomonas jessenii*, *P. protegens, Bacillus thuringiensis* and *Serratia plymuthica*. Additionally, some fungal isolates for example *Purpureocillium lilacinus, Trichoderma harzianum, Arthrobotrys oligospora, Lecanicillium muscarium Gliocladium* spp., *Pochonia chlamydosporia* and *Paecillomyces lilacinus* [54–56]. Furthermore, some endophytic agents such as *Fusarium oxysporum* (FO162) can induce systemic resistance against *Meloidogyne* spp. in tomato [57] and plant growth-promoting rhizobacteria (PGPR) as well [58]. One of the most important PGPR is strain LMG27872 of *Paenibacillus polymyxa* that increasing the percentage of J2 mortality and reducing number of galls and egg hatching of *M. incognita* in tomato [59]. Similarly, the same effect detected by two bacterial isolates ZHA296 and ZHA178 of *Paenibacillus castaneae* [60]. Also, both bacterial strains; BZR 86 and BZR 277 of *Bacillus velezensis* inhibited *M. incognita* and improved plant health, and crop productivity under greenhouse condition [61]. Moreover, *Bacills amyloquefaciens*, *B. megaterium*, *Pseudomonas fluorescens*, and *P. putida* have a potential effect against RKN in the laboratory as well as in field conditions [62]. Among endophytic fungi, Arbuscular mycorrhizal fungi (AMF) that are soil fungi and symbiosis with the plant roots. These fungi extremely benefit for plant healthy by acting enhanced plant tolerance for RKN due to induce plant systemic resistance and provide plant nutrients [63, 64].

6.1.4 Resistant genotypes

Planting resistant cultivars is one of the environmentally friendly methods to reduce RKN in tomato. The plant resistance for RKN depends on the genotypes that are restrict or prevent nematode reproduction in their plant host. At least 10 plant resistance genes (*R*-genes; *Mi-1, Mi-2, Mi-3, Mi-4, Mi-5, Mi-6, Mi-7, Mi-8, Mi-9*, and *Mi-HT*) for *Meloidogyne* spp. have been identified in tomato plant [65] Among these genes, only five genes (*Mi-1, Mi-3, Mi-5, Mi-9,* and *Mi-HT*) have been mapped. Concerning tomato-resistant genotypes, *Mi-1* gene is the most common that was originally identified in *Solanum peruvianum* and transferred into *S. lycopersicum* [66]. This gene confers resistance to *M. incognita*, *M. javanica*, and *M. arenaria* [44, 66]. The Gene map of *Mi-1* gene localized it to the short arm of tomato chromosome 6 [67]. There are two homologs of *Mi-1* gene coded by *Mi-1.1* and *Mi-1.2* that were identified at the *Mi* locus. The *Mi-1.2* gene conferred resistance to 15 populations of *Meloidogyne* spp. [68]. In contrast, the tomato genotypes that are possess *Mi-1.1* gene but lack the *Mi1.2* gene demonstrated highly compatible with *M. javanica* [69, 70]. While the resistant genotypes that possess *Mi1.2* gene can delay or suppress the development and reproduction of nematodes [70]. In addition, 83 *WRKY* genes have identified in tomato plants [71]. One or more members of this gene family such as *SlWRKY72*, *SlWRKY73*, or *SlWRKY74* have been examined as contributing positively to both PAMP-triggered immunity (PTI) and *Mi-1*-mediated effector-triggered immunity (ETI) against *M. javanica* [72, 73]. Also, the *SlWRKY80* gene was required for *Mi-1*-mediated resistance against RKN [74]. Thus, these genes could play an important role during nematode infection in investigated resistant tomato genotypes as *Mi-1*-mediated effector-triggered immunity. Subsequently, all these sources of resistance can become valuable additions to nematode management strategies in the future considering that the natural resistance require long-term experiments. Also, genetic resistance in the host plant could be breakdown when there is a high population density of the nematode [75].

6.2 Innovative methods

The chemical activation of the plant's natural defense mechanisms should be involved as alternative safety strategy for management of RKN. Some chemicals are inducers challenging localized hypersensitive reactive which involves recognition proceedings between plant and pathogen. In systemic manner, plants have other mechanisms that boundary pathogen access and their reproduction. For instance, several defense genes in plant up regulated by salicylic acid and Benzothiadiazole [76–78]. The defensive proteins that called pathogenesis-related proteins expressed by these genes in resistance or tolerance plant. Chemical induction of "systemic acquired resistance" is detected by using Benzothiadiazole in tomato and grapevines plants to suppress infection of *M. incognita* [79] and by using hydroxyurea in tomato to inhibit progress infection of *M. javanica* [80]. Also, chitosan stimulated production of defense-related chemicals in tomato plant and was associated through improvement process of resistance to root-knot nematode as well [81]. In addition, some botanical extracts and synthetic compounds able to be a resistance inducer to plant pathogen [82, 83]. Many chemicals able to encourage systemic resistance in various plant species to different pathogens such as Oomycetes, fungi, bacteria, viruses, and nematodes [84, 85]. During plant defense mechanisms, Jasmonic acid (JA) and salicylic acid (SA) play an important role as plant growth regulator [86] and shown different reactions in plant resistance responses against root-knot nematodes [87–91]. As a reaction to biotic and abiotic agent, γ-Aminobutyric Acid (GABA) is a non-protein amino acid rapidly accumulated within tomato plant tissues and has an important role in plants during plant-pathogen interaction particularly RKN [92]. Also, GABA has an isomer named β-Aminobutyric Acid (BABA) that is known as an inducer for plant disease resistance (ISR) when applied to various plants host during RKN infection [93, 94]. Also, BABA plays a role as an inducer for Systemic Acquired Resistance (SAR) against *M. javanica* [95]. Consequently, treatment of tomato plants by BABA reduced damage of root knot nematode. Also, suppress *M. javanica* on pineapple, and exposed induce resistance against *M. javanica* in cucumber [95]. Among commercial products, both Agrispon and Sincocin are liquid concentrate derived from plant extracts. Commercially, Agrispon sold as a plant fertilizer that led to improve root building, plant growth and their yield without a negative environmental impact. Also, Sincocin used in plant fertility programs and improves plant's ability to resist a variety of pathogens and environmental stresses.

6.3 Novel approaches

Among advanced techniques, microarray analysis technique that can add more detail for about plant response to *Meloidogyne* spp. infection through identification of some genes involved in the pathogenicity relationship between RKN and their plant host. Specifically, genes implicated in cell wall formation, transport processes and plant defense responses during NFS and GC formation. As well, the development of RNA interference (RNAi) and complementary DNA (cDNA) technology led to characterized nematode secretions as parasitism effectors and should explain the molecular events and regulatory mechanisms during RKN infection. Consequently, it is easy to be following the proteins that are expressed by these target genes in NFS via immunolocalization analysis either *In vitro* by using *In situ* hybridization or immunofluorescence technique, or *In vivo* using green fluorescence protein (GFP) fused with target gene. Among manipulation occurred in GCs at the cytoskeletal level, the previous studies observed that the disruption of the cytoskeleton is possibly a requirement to allow RKN to complete their life cycle [96]. Thus, the cytoskeleton is involved in the process of RKN infection. The genetical studies for *Arabidopsis thaliana* genome identified seven actin-depolymerizing factor (*ADF*) genes that are upregulated in GC during infection of by RKN. Particularly the expression of *ADF2* gene increased between 14 and 21 d after RKN inoculation resulting in accumulation of actin filaments in GC. The knockdown of *ADF2* gene using RNAi reveals that ADF2 protein expressed by this gene is required for normal cell development and plant growth. Thus, during nematode infection, decreasing level of ADF2 protein led to reducing F-actin turnover and inhibition the expansion of GCs in NFS and gall formation as well. Accordingly, these effects, the development of RKN is delay and their reproduction significantly decreasing [35]. Recently, several cytoskeleton-associated proteins facilitating cytoskeletal remodeling and defense signaling findings have discovered. Furthermore, the reorganization of the actin cytoskeleton is revealed to further feedback-regulate reactive oxygen species (ROS) production and trigger salicylic acid (SA) signaling [97]. Beside actine filaments, microtubules are a member in the cytoskeleton network and have also an important role in mitotic activity during cell division. Regarding *A. thaliana*, there are two genes; *TUBG1* and *TUBG2* which are nucleate the cortical cytoplasm microtubules and regulate their dynamic. Furthermore, *GCP3* and *GCP4* genes which are nucleate the mitotic microtubules during cell division. During RKN infection, these four genes are upregulating in the cytoplasm and at cell wall of GC causing increasing of microtubule nucleation in cytoplasm and forming a complex network of cortical microtubules at the cell wall of GC and become randomly organized. Without affecting of un-infected plant growth, the knockout of either or both *TUBG1* and *TUBG2* genes let to delaying of nematode life cycle and gall formation, and decreasing their population [36, 98]. Based on these studies that focused on cytoskeleton and its coordinating genes during RKN infection, could be cytoskeleton has an extremely important role of the plant immunity and plant susceptibility for nematode infection as well [99]. Concerning nematode secretions that containing effectors destroy the cytoskeleton, *Meloidogyne incognita* secrete an effector called MiPFN3 (*Meloidogyne incognita* Profilin 3). This protein effector can bind to actin monomers, disrupt actin polymerization, and reduce the filamentous actin network [100]. That is means that the manipulation of the cytoskeleton by RKN may be a tactic to promote its parasitism. Additionally, *Arabidopsis* genome comprising two pectate lyase-like genes (PLL), *PLL18* (At3g27400) and *PLL19* (At4g24780). Upregulation of these genes detected in the developed NFS during infection of *M. incognita*. While the mutant lines that loss one of these genes negatively influences the development of GC [101]. Similarly, Arabidopsis histidine kinase receptor mutant lines ahk2/3, ahk2/4 and ahk3/4 revealed less susceptible to RKN suggesting a requirement of cytokinin signaling for GC and feeding site formation [102]. According to the endoreduplication of DNA is required for the mitotic activity during GC formation. Therefore, potential DNA damage in the genome of gall cells is evident. *WEE1* gene is particularly involved in DNA damage checkpoint control and encoding for a protein kinase that controls cell cycle [103]. Nematode feeding site demonstrate transcriptional activation of the DNA damage checkpoint kinase WEE1. The interrupted nuclei phenotype in GC indicated to the accumulation of mitotic defects. The WEE1-knockout line in *Arabidopsis* and WEE1-knockdowen line downregulation in tomato repressed RKN infection and their reproduction [104].

7. Conclusion

The highly impacting of root-knot nematode (*Meloidogyne* spp.) in tomato and other plant host requires an urgent integration management for this plant pathogen. Since the toxicity of nematicides is the critical point for our environmental safety. The alternative eco-friendly chemicals that have a nematocidal effect must be applied together with the application of agronomic and biological control methods, and cultivation of resistance genotypes in strategy of integrated pest management. Genetic analysis of *Meloidogyne* spp. identified genes that encoding specific effectors promoting parasitism process modulating the plant's defense system to successfully forming and establishment of a nematode feeding site (NFS) [105]. Consequently, the functional analyses of these effector could be led to the identification of susceptibility genes with potential for use in resistance breeding [106, 107]. However, these susceptibility genes mostly have a vital role for plant physiology and development. Interfering with host protein recognition by nematode effectors may be an interesting way of preserving important plant functions whilst breaking the susceptibility of the plant to nematode. The breeding of new line harboring mutations that are less susceptible to nematode infection may be achieved with new technologies, such as the TILLING and CRISPR/Cas9 technologies [108, 109]. According to the knowledge for about the functions of effector/target which are required to improve the compatibility between nematode and their plant host, it can guide this strategy to stop this interaction and engineer durable disease resistance as a novel process to root-knot nematode management.

Author details

Mohamed Youssef Banora
Faculty of Agriculture, Department of Plant Pathology, Ain Shams University, Egypt

*Address all correspondence to: youssef_myb@agr.asu.edu.eg

References

[1] Moens M, Perry RN, Starr J. *Meloidogyne* species a diverse group of novel and important plant parasites. In: Perry RN, Moens M, Starr JL, editors. Root-knot Nematodes. Cambridge: CABI Publications; 2009. pp. 1-17. DOI: 10.1079/9781845934927.0001

[2] Hunt DJ, Handoo ZA. Taxonomy, identification and principal species. In: Perry RN, Moens M, Starr JR, editors. Root-knot Nematodes. Cambridge: CABI Publications; 2009. pp. 55-97. DOI: 10.1079/9781845934927.0055

[3] Karssen G, Van Aelst A. Root-knot nematode perineal pattern development: A reconsideration. Nematology. 2001;**3**:95-111. DOI: 10.1163/156854101750236231

[4] De Waele D, Elsen A. Challenges in tropical plant nematology. Annual Review of Phytopathology. 2007;**45**:457-485. DOI: 10.1146/annurev.phyto.45.062806.094438

[5] Perry RN, Moens M. Plant Nematology. Wallingford: CABI Publications; 2006. p. 447. DOI: 10.1079/9781845930561.0000

[6] Nicol J, Turner S, Coyne D, den Nijs L, Hockland S, Maafi ZT. Current nematode threats to world agriculture. In: Jones J, Gheysen G, Fenoll C, editors. Genomics and Molecular Genetics of Plant-Nematode Interactions. Berlin: Springer; 2011. pp. 21-43. DOI: 10.1007/978-94-007-0434-3_2

[7] Elling AA. Major emerging problems with minor *Meloidogyne* species. Phytopathology. 2013;**103**:1092-1102. DOI: 10.1094/PHYTO-01-13-0019-RVW

[8] Jones JT, Haegeman A, Danchin EGJ, Gaur HS, Helder J, Jones MGK, et al. Top 10 plant-parasitic nematodes in molecular plant pathology. Molecular Plant Pathology. 2013;**14**:946-961. DOI: 10.1111/mpp.12057

[9] Khan MW. Mechanisms of interactions between nematodes and other plant pathogens. In: Khan MW, editor. Nematode Interactions. London: Chapman & Hall; 1993. pp. 55-78. DOI: 10.1007/978-94-011-1488-2_4

[10] Manzanilla-López RH, Starr JL. Interactions with other pathogens. meloidogyne species a diverse group of novel and important plant parasites. In: Perry RN, Moens M, Starr JL, editors. Root-knot Nematodes. Cambridge: CABI Publications; 2009. pp. 223-245. DOI: 10.1079/9781845934927.0223

[11] Food and Agriculture Organization (FAO). Statistical databases. In: Statistics Division. Rome: Food and Agriculture Organization the United Nations; 2022. DOI: 10.4060/cc2211en

[12] Sasser JN, Eisenback JD, Carter CC, Triantaphyllou AC. The international *Meloidogyne* project - Its goals and accomplishments. Annual Review of Phytopathology. 1983;**21**:271-288. DOI: 10.1146/annurev.py.21.090183.001415

[13] Salazar-Antón W, Guzmán-Hernández TDJ. Effect of populations of *Meloidogyne* spp. in the development and yield of the tomato. Agronomía Mesoamericana. 2013;**24**:419-426. DOI: 10.15517/am.v24i2.12542

[14] Khan MTA, Mukhtar T, Saeed M. Resistance or susceptibility of eight aubergine cultivars to *Meloidogyne javanica*. Pakistan Journal of Zoology. 2019;**51**:2187-2192. DOI: 10.17582/journal.pjz/2019.51.6.2187.2192

[15] Greco DV. Population dynamics and damage levels. In: Perry RN, Moens M, Starr JR, editors. Root-knot Nematodes. UK: CABI Publishing; 2009. pp. 246-274. DOI: 10.1079/9781845934927.0246

[16] Rodiuc N, Vieira P, Banora MY, Engler JA. On the track of transfer cell formation by specialized plant parasitic nematodes. Frontiers in Plant Science. 2014;**5**:160. DOI: 10.3389/fpls.2014.00160

[17] Bird DM, Koltai H. Plant parasitic nematodes: Habitats, hormones, and horizontally-acquired genes. Journal of Plant Growth Regulation. 2000;**19**:183-194. DOI: 10.1007/s003440000022

[18] Sparks TC. Insecticide discovery: An evaluation and analysis. Pesticide Biochemistry and Physiology. 2013;**107**:8-17. DOI: 10.1016/j.pestbp.2013.05.012

[19] Wyss U, Grundler FMW, Munch A. The parasitic behavior of second-stage juveniles of *Meloidogyne incognita* in roots of *Arabidopsis thaliana*. Nematologica. 1992;**38**:98-111. DOI: 10.1163/187529292X00081

[20] Rosso M, Hussey R, Davis EL, Smant G, Baum T, Abad P, et al. Nematode effector proteins: Targets and functions in plant parasitism. In: Martin F, Kamoun S, editors. Effectors in Plant-Microbe Interactions. New York: John Wiley & Sons; 2011. pp. 327-354. DOI: 10.1002/9781119949138.ch13

[21] Vieira P, Danchin EGJ, Neveu C, Crozat C, Juabert S, Hussey RS, et al. The plant apoplasm is an important recipient compartment for nematode secreted proteins. Journal of Experimental Botany. 2011;**62**:1241-1253. DOI: 10.1093/jxb/erq352

[22] Paulo V, Banora MY, Castagnone-Sereno P, Rosso MN, Engler G, de Almeida-Engler J. Immunolocalization procedure for protein localization in nematode pre-parasitic and parasitic stages using methylacrylate-embedded tissues. Phytopathology. 2012;**102**:990-996. DOI: 10.1094/PHYTO-02-12-0031-R

[23] Wyss U, Grundler FMW. Feeding behaviour of sedentary plant-parasitic nematodes. Netherlands Journal of Plant Pathology. 1992;**98**:165-173. DOI: 10.1007/BF01974483

[24] Sijmons PC, Atkinson HJ, Wyss U. Parasitic strategies of root nematodes and associated host cell responses. Annual Review of Phytopathology. 1994;**32**:235-259. DOI: 10.1146/annurev.py.32.090194.001315

[25] Sijmons PC. Plant–nematode interactions. Plant Molecular Biology. 1993;**23**:917-931. DOI: 10.1007/BF00021809

[26] Triantaphyllou AC. Environmental sex differentiation of nematodes in relation to pest management. Annual Review of Phytopathology. 1973;**11**:441-462. DOI: 10.1146/annurev.py.11.090173.002301

[27] Nguyen CN, Perfus-Barbeoch L, Quentin M, Zhao J, Magliano M, Marteu N, et al. A root-knot nematode small glycine and cysteine-rich secreted effector, MiSGCR1, is involved in plant parasitism. The New Phytologist. 2018;**217**:687-699. DOI: 10.1111/nph.14837

[28] Wang XR, Moreno YA, Wu HR, Ma C, Li YF, Zhang JA, et al. Proteomic profiles of soluble proteins from the esophageal gland in female Meloidogyne incognita. International Journal for Parasitology. 2012;**42**:1177-1183. DOI: 10.1016/j.ijpara.2012.10.008

[29] Siddique S, Grundler FM. Parasitic nematodes manipulate plant

development to establish feeding sites. Current Opinion in Microbiology. 2018;**46**:102-108. DOI: 10.1016/j.mib.2018.09.004

[30] Quentin M, Abad P, Favery B. Plant parasitic nematode effectors target host defense and nuclear functions to establish feeding cells. Frontiers in Plant Science. 2013;**4**:53. DOI: 10.3389/fpls.2013.00053

[31] Goverse A, Smant G. The activation and suppression of plant innate immunity by parasitic nematodes. Annual Review of Phytopathology. 2014;**52**:243-265. DOI: 10.1146/annurev-phyto-102313-050118

[32] Kyndt T, Vieira P, Gheysen G, de Almeida-Engler J. Nematode feeding sites: Unique organs in plant roots. Planta. 2013;**238**:807-818. DOI: 10.1007/s00425-013-1923-z

[33] Favery B, Quentin M, Jaubert-Possamai S, Abad P. Gall-forming root-knot nematodes hijack key plant cellular functions to induce multinucleate and hypertrophied feeding cells. Journal of Insect Physiology. 2016;**84**:60-69. DOI: 10.1016/j. jinsphys.2015.07.013

[34] Palomares-Rius JE, Escobar C, Cabrera J, Vovlas A, Castillo P. Anatomical alterations in plant tissues induced by plant-parasitic nematodes. Frontiers in Plant Science. 2017;**8**:1987. DOI: 10.3389/fpls.2017.01987

[35] Clement M, Ketelaar T, Rodiuc N, Banora SA, Engler G, Abad P, et al. ADF-dependent actin cytoskeleton remodelling is essential for plant parasitic nematode infection. The Plant Cell. 2009;**21**:2963-2979. DOI: 10.1105/tpc.109.069104

[36] Banora MY, Rodiuc N, Baldacci-Cresp F, Smertenko A, Bleve-Zacheo T, Mellilo MT, et al. Feeding cells induced by Phytoparasitic nematodes require γ-tubulin ring complex for microtubule reorganization. PLoS Pathogens. 2011;**7**:e1002343. DOI: 10.1371/journal.ppat.1002343

[37] Mejias J, Truong NM, Abad P, Favery B, Quentin M. Plant proteins and processes targeted by parasitic nematode effectors Joffrey. Frontiers in Plant Science. 2019;**10**:970. DOI: 10.3389/fpls.2019.00970

[38] Ramesh C, Subhash C, Vimal SC, Shyam G, Dhirendra S, Dhirendra P. Impact assessment of neem products in the control of root-knot nematode, *Meloidogyne* incognita on Brinjal. Scientist. 2022;**3**:496-501. DOI: 10.5281/zenodo.7135269

[39] Ji X, Li J, Meng Z, Dong S, Zhang S, Qiao K. Inhibitory effect of allicin against *Meloidogyne incognita* and Botrytis cinerea in tomato. Scientia Horticulturae. 2019;**253**:203-208. DOI: 10.1016/j.scienta.2019.04.046

[40] Blok VC, Jones JT, Phillips MS, Trudgill DL. Parasitism genes and host range disparities in biotrophic nematodes: The conundrum of polyphagy versus specialisation. BioEssays. 2008;**30**:249-259. DOI: 10.1002/bies.20717

[41] de Brita AL, de Castrro BM, Zanuncio JC, Serrao JE, Wilcken SRS. Oat, wheat and sorghum cultivars for the management of *Meloidogyne enterolobii*. Nematology. 2018;**20**:169-173. DOI: 10.1163/15685411-00003131

[42] Schwarz T, Li C, Ye W, Davis E. Distribution of *Meloidogyne enterolobii* in eastern North Carolina and comparison of four isolates. Plant Health Progress. 2020;**21**:91-96. DOI: 10.1094/php-12-19-0093-rs

[43] Talavera M, Verdejo-Lucas S, Ornat C, Torres J, Vela MD, Macias FJ, et al. Crop rotations with *mi* gene resistant and susceptible tomato cultivars for management of root-knot nematodes in plastic houses. Crop Protection. 2009;**28**:662-667. DOI: 10.1016/j.cropro.2009.03.015

[44] Seid A, Fininsa C, Mekete T, Decraemer W, Wesemael WML. Tomato (*Solanum lycopersicum*) and root-knot nematodes (*Meloidogyne* spp.) - A century-old battle. Nematology. 2015;**17**:995-1009. DOI: 10.1163/15685411- 00002935

[45] Bélair G, Benoit DL. Host suitability of 32 common weeds to meloidogyne hapla in organic soils of Southwestern Quebec. Journal of Nematology. 1996;**28**:643-647

[46] Rich JR, Brito JA, Kaur R, Ferrell JA. Weed species as hosts of *Meloidogyne*: A review. Nematropica. 2009;**39**:157-185

[47] Sikora RA, Roberts PA. Management practices: An overview of integrated nematode management technologies. In: Sikora RA, Coyne D, Hallmann J, Timper P, editors. Plant Parasitic Nematodes in Subtropical and Tropical Agriculture. Wallingford: CABI; 2018. pp. 795-838. DOI: 10.1079/9781786391247.0795

[48] Ioannou N. Soil solarization as a substitute for methyl bromide fumigation in greenhouse tomato production in Cyprus. Phytoparasitica. 2000;**28**:248-256. DOI: 10.1007/BF02981803

[49] Nico AI, Jimenez-Diaz RM, Castillo P. Solarization of soil in piles for the control of *Meloidogyne incognita* in olive nurseries in southern Spain. Plant Pathology. 2003;**52**:770-778. DOI: 10.1111/j.1365-3059.2003.00927.x

[50] McSorley R, Duncan LW. 8 Economic thresholds and nematode management. Advances in Plant Pathology. 1995;**11**:147-162. DOI: 10.1016/S0736-4539(06)80010-3

[51] Putten WHVD, Cook R, Costa S, Davies KG, Fargette M, Freitas H, et al. Nematode interactions in nature: Models for sustainable control of nematode pests of crop plants? Advances in Agronomy. 2006;**89**:227-260. DOI: 10.1016/S0065-2113(05)89005-4

[52] Dutta TK, Khan MR, Phani V. Plant-parasitic nematode management via biofumigation using brassica and non-brassica plants: Current status and future prospects. Current Plant Biology. 2019;**17**:17-32. DOI: 10.1016/j.cpb.2019.02.001

[53] Forghani F, Hajihassani A. Recent advances in the development of environmentally benign treatments to control root-knot nematodes. Frontiers in Plant Science. 2020;**11**:1125. DOI: 10.3389/fpls.2020.01125

[54] Abd-Elgawad MMM. Optimizing safe approaches to manage plant-parasitic nematodes. Plants. 2021;**10**:1911. DOI: 10.3390/plants10091911

[55] Azlay L, El Boukhari ME, El Hassan M, Mustapha B. Biological management of root-knot nematodes (*Meloidogyne* spp.): A review. Organic Agriculture. 2023;**13**:99-117. DOI: 10.1007/s13165-022-00417-y

[56] Bhat AA, Shakeel A, Waqar S, Handoo ZA, Khan AA. Microbes vs. nematodes: Insights into biocontrol through antagonistic organisms to control Root-knot nematodes. Plants. 2023;**12**:451. DOI: 10.3390/plants12030451

[57] Walters DR. Are plants in the field already induced? Implications

for practical disease control. Crop Protection. 2009;**28**:459-465. DOI: 10.1016/j.cropro.2009.01.009

[58] Aioub AAA, Elesawy AE, Ammar EE. Plant growth promoting rhizobacteria (PGPR) and their role in plant-parasitic nematodes control: A fresh look at an old issue. Journal of Plant Diseases and Protection. 2022;**129**:1305-1321. DOI: 10.1007/s41348-022-00642-3

[59] Singh RR, Wesemael WML. Endophytic *Paenibacillus polymyxa* LMG27872 inhibits *Meloidogyne incognita* parasitism, promoting tomato growth through a dose-dependent effect. Frontiers in Plant Science. 2022;**13**:961085. DOI: 10.3389/fpls.2022.961085

[60] Cetintas R, Kusek M, Fateh S. Effect of some plant growth-promoting rhizobacteria strains on root-knot nematode, *Meloidogyne incognita*, on tomatoes. Egyptian Journal of Biological Pest Control. 2018;**28**:1-5. DOI: 10.1186/s41938-017-0008-x

[61] Asaturova AM, Bugaeva LN, Homyak AI, Slobodyanyuk GA, Kashutina EV, Yasyuk LV, et al. *Bacillus velezensis* strains for protecting cucumber plants from Root-knot nematode *Meloidogyne incognita* in a greenhouse. Plants. 2022;**11**:275. DOI: 10.3390/plants11030275

[62] Rani P, Singh M, Prashad H, Sharma M. Evaluation of bacterial formulations as potential biocontrol agents against the southern root-knot nematode, *Meloidogyne incognita*. Egyptian Journal of Biological Pest Control. 2022;**32**:29. DOI: 10.1186/s41938-022-00529-3

[63] Baum C, El-Tohamy W, Gruda N. Increasing the productivity and product quality of vegetable crops using arbuscular mycorrhizal fungi: A review. Scientia Horticulturae. 2015;**187**:131-141. DOI: 10.1016/j.scienta.2015.03.002

[64] Schouteden N, Waele DD, Panis B, Vos CM. Arbuscular mycorrhizal fungi for the biocontrol of plant-parasitic nematodes: A review of the mechanisms involved. Frontiers in Microbiology. 2015;**6**:1280. DOI: 10.3389/fmicb.2015.01280

[65] El-Sappah AH, Islam MM, El-Awady HH, Yan S, Qi S, Liu J, et al. Tomato natural resistance genes in controlling the root-knot nematode. Genes. 2019;**10**:925. DOI: 10.3390/genes10110925

[66] Da Silva AJ, de Oliveira GHF, PastorizaRJG,MaranhaoEHA,PedrosaEMR, Maranhao SRVL, et al. Search for sources of resistance to *Meloidogyne enterolobii* in commercial and wild tomatoes. Horticultura Brasileira. 2019;**37**:188-198. DOI: 10.1590/s0102-053620190209

[67] Kaloshian I, Yaghoobi J, Liharska T, Hontelez J, Hanson D. Genetic and physical localization of the root-knot nematode-resistance locus *mi* in tomato. Molecular and General Genetics. 1998;**257**:376-385. DOI: 10.1007/s004380050660

[68] Gabriel M, Kulczynski SM, Muniz MFB, Boiteux LS, Carneiro RMDG. Resistance of 'Debora Plus' tomato bearing *Mi-1.2* gene/locus against fifteen *Meloidogyne* species. Plant Pathology. 2020;**69**:944-952. DOI: 10.1111/ppa.13179

[69] Hwang CF, Bhakta AV, Truesdell GM, Pudlo WM, Williamson VM. Evidence for a role of the N terminus and leucine-rich repeat region of the *mi* gene product in regulation of localized cell death. The Plant Cell. 2000;**12**:1319-1329. DOI: 10.1105/tpc.12.8.1319

[70] Banora MY, Almaghrabi OA. Differential response of some nematode-resistant and susceptible tomato genotypes to *Meloidogyne javanica* infection. Journal of Plant Protection Research. 2019;**59**:113-123. DOI: 10.24425/jppr.2019.126040

[71] Karkute SG, Gujjar RS, Rai A, Akhtar M, Singh M. Genome wide expression analysis of *WRKY* genes in tomato (*Solanum lycopersicum*) under drought stress. Plant Gene. 2018;**13**:8-17. DOI: 10.1016/j.plgene.2017.11.002

[72] Bhattarai KK, Atamian HS, Kaloshian I, Eulgem T. *WRKY72*-type transcription factors contribute to basal immunity in tomato and Arabidopsis as well as gene-for-gene resistance mediated by the tomato *R* gene *Mi-1*. The Plant Journal. 2010;**63**:229-240. DOI: 10.1111/j.1365-313X.2010.04232.x

[73] Atamian HS, Eulgem T, Kaloshian I. *SlWRKY70* is required for *Mi1*-mediated resistance to aphids and nematodes in tomato. Planta. 2012;**235**:299-309. DOI: 10.1007/s00425-011-1509-6

[74] Bai Y, Sunarti S, Kissoudis C, Visser RGF, van der Linden CG. The role of tomato *WRKY* genes in plant responses to combined abiotic and biotic stresses. Frontiers in Plant Science. 2018;**9**:801. DOI: 10.3389/fpls.2018.00801

[75] Padilla-Hurtado B, Morillo-Coronado Y, Tarapues S, Burbano S, Soto-Suárez M, Urrea R, et al. Evaluation of root-knot nematodes (*Meloidogyne* spp.) population density for disease resistance screening of tomato germplasm carrying the gene *Mi-1*. Chilean Journal of Agricultural Research. 2022;**82**:157-166. DOI: 10.4067/S0718-58392022000100157

[76] Conrath U, Chen Z, Ricigliano JR, Klessig DF. Two inducers of plant defense responses, 2,6-dichloroisonicotinec acid and salicylic acid, inhibit catalase activity in tobacco. Proceedings. National Academy of Sciences. United States of America. 1995;**92**:7143-7147. DOI: 10.1073/pnas.92.16.7143

[77] Klessig DF, Choi HW, D'MA D. Systemic acquired resistance and salicylic acid: Past, present, and future. Molecular Plant-Microbe Interactions. 2018;**31**:871-888. DOI: 10.1094/MPMI-03-18-0067-CR

[78] Elkobrosy DH, Abougabal AA, Abdelsalam NR, Mohamed RA, Zeid A. Enhancing tomato cultivars against Root-knot nematode using salicylic acid and their impact on protein expression. Egyptian Academic Journal of Biological Sciences. 2022;**13**:165-170. DOI: 10.21608/EAJBSH.2022.275938

[79] Gulzar RMA, Rehman AU, Umar UUD, Shahid M, Khan MF. Evaluation of genetic and induced resistance phenomena in cucumbers against the root-knot nematode (*Meloidogyne incognita*). Plant Protection Science. 2022;**58**:338-350. DOI: 10.17221/130/2021-PPS

[80] Asadi-Sardari A, Mahdikhani-Moghadam E, Zaki-Aghl M, Vetukuri RR. Constitutive and inducible expression of genes related to salicylic acid and ethylene pathways in a moderately resistant tomato cultivar leads to delayed development of *Meloidogyne javanica*. Agriculture. 2022;**12**:2122. DOI: 10.3390/agriculture12122122

[81] Khalil MS, Abd El-Aziz MH, Selim RES. Physiological and morphological response of tomato plants to nano-chitosan used against bio-stress induced by root-knot nematode (*Meloidogyne incognita*) and tobacco mosaic tobamovirus (TMV). European Journal of Plant Pathology.

2022;**163**:799-812. DOI: 10.1007/s10658-022-02516-8

[82] Walters D, Walsh D, Newton A, Lyon G. Induced resistance for plant disease control: Maximizing the efficacy of resistance elicitors. Phytopathology. 2005;**95**:1368-1373. DOI: 10.1094/PHYTO-95-1368

[83] Elsharkawy MM, Al-Askar AA, Behiry SI, Abdelkhalek A, Saleem MH, Kamran M, et al. Resistance induction and nematicidal activity of certain monoterpenes against tomato root-knot caused by *Meloidogyne incognita*. Frontiers in Plant Science. 2022;**13**:982414. DOI: 10.3389/fpls.2022.982414

[84] Yang B, Yang S, Zheng W, Wang Y. Plant immunity inducers: From discovery to agricultural application. Stress Biology. 2022;**2**:5. DOI: 10.1007/s44154-021-00028-9

[85] Zhu F, Cao MY, Zhang QP, Mohan R, Schar J, Mitchell M, et al. Join the green team: Inducers of plant immunity in the plant disease sustainable control toolbox. Journal of Advanced Research. 2023;**2**:S2090-1232(23)00122-4. DOI: 10.1016/j.jare.2023.04.016

[86] Vlot AC, Dempsey DA, Klessig DF. Salicylic acid, a multifaceted hormone to combat disease. Annual Review of Phytopathology. 2009;**47**:177-206. DOI: 10.1146/annurev.phyto.050908.135202

[87] Molinaria S, Loffredo E. The role of salicylic acid in defense response of tomato to root-knot nematodes. Physiological and Molecular Plant Pathology. 2006;**68**:69-78. DOI: 10.1016/j.pmpp.2006.07.001

[88] Bhattarai KK, Xie QG, Mantelin S, Bishnoi U, Girke T, Navarre DA, et al. Tomato susceptibility to root-knot nematodes requires an intact jasmonic acid signaling pathway. Molecular Plant-Microbe Interactions. 2008;**21**:1205-1214. DOI: 10.1094/MPMI-21-9-1205

[89] Molinari S, Fanelli E, Leonetti P. Expression of tomato salicylic acid (SA)-responsive pathogenesis-related genes in *Mi-1*-mediated and SA-induced resistance to root-knot nematodes. Molecular Plant Pathology. 2014;**15**:255-264. DOI: 10.1111/mpp.12085

[90] Martínez-Medina A, Fernandez I, Lok GB, Pozo MJ, Pieterse CM, Van Wees SC. Shifting from priming of salicylic acid- to jasmonic acid-regulated defences by *Trichoderma* protects tomato against the root knot nematode *Meloidogyne incognita*. The New Phytologist. 2017;**213**:1363-1377. DOI: 10.1111/nph.14251

[91] Yang Y-X, Wu C, Ahammed GJ, Wu C, Yang Z, Wan C, et al. Red light-induced systemic resistance against root-knot nematode is mediated by a coordinated regulation of salicylic acid, Jasmonic acid and redox Signaling in watermelon. Frontiers in Plant Science. 2018;**9**:899. DOI: 10.3389/fpls.2018.00899

[92] Saito T, Matsukura C, Sugiyama M, Watahiki A, Ohshima I, Iijima Y, et al. Screening for γ-aminobutyric acid (GABA)-rich tomato varieties. Journal of the Japanese Society for Horticultural Science. 2008;**77**:242-250. DOI: 10.2503/jjshs1.77.242

[93] Sahebani N, Hadavi NS, Zade FO. The effects of β-amino-butyric acid on resistance of cucumber against root-knot nematode, *Meloidogyne javanica*. Acta Physiologiae Plantarum. 2011;**33**:443-450. DOI: 10.1007/s11738-010-0564-0

[94] Ji H, Kyndt T, He W, Vanholme B, Gheysen G. β-Aminobutyric acid–induced resistance against root-knot

nematodes in rice is based on increased basal defense. Molecular Plant-Microbe Interactions. 2015;**28**:519-533. DOI: 10.1094/MPMI-09-14-0260-R

[95] Taher IE, Ami SN. Inducing systemic acquired resistance (SAR) against Root-knot nematode Meloidogyne Javanica and evaluation of biochemical changes in cucumber Root. Helminthologia. 2022;**59**:404-413. DOI: 10.2478/helm-2022-0042

[96] de Almeida Engler J, Van Poucke K, Karimi M, De Groodt R, Gheysen G, Engler G, et al. Dynamic cytoskeleton rearrangements in giant cells and syncytia of nematode-infected roots. The Plant Journal. 2004;**38**:12-26. DOI: 10.1111/j.1365-313X.2004.02019.x

[97] Wang J, Lian N, Zhang Y, Man Y, Chen L, Yang H, et al. The cytoskeleton in plant immunity: Dynamics, regulation, and function. International Journal of Molecular Sciences. 2022;**23**:15553. DOI: 10.3390/ijms232415553

[98] Hardham AR. Microtubules and biotic interactions. The Plant Journal. 2013;**75**:278-289. DOI: 10.1111/tpj.12171

[99] Li J, Staiger CJ. Understanding cytoskeletal dynamics during the plant immune response. Annual Review of Phytopathology. 2018;**56**:513-533. DOI: 10.1146/annurev-phyto-080516-035632

[100] Leelarasamee N, Zhang L, Gleason C. The root-knot nematode effector MiPFN3 disrupts plant actin filaments and promotes parasitism. PLoS Pathogens. 2018;**14**:e1006947. DOI: 10.1371/journal.ppat.1006947

[101] Wieczorek K, Elashry A, Quentin M, Grundler FMW, Favery B, Seifert GJ, et al. A distinct role of pectate lyases in the formation of feeding structures induced by cyst and Root-knot nematodes. Molecular Plant-Microbe Interactions. 2014;**27**:901-912. DOI: 10.1094/MPMI-01-14-0005-R

[102] Dowd CD, Chronis D, Radakovic ZS, Siddique S, Schmulling T, Werner T, et al. Divergent expression of cytokinin biosynthesis, signaling and catabolism genes underlying differences in feeding sites induced by cyst and root-knot nematodes. The Plant Journal. 2017;**92**:211-228. DOI: 10.1111/tpj.13647

[103] De Schutter K, Joubès J, Cools T, Verkest A, Corellou F, Babiychuk E, et al. *Arabidopsis* WEE1 kinase controls cell cycle arrest in response to activation of the DNA integrity checkpoint. The Plant Cell. 2007;**19**:211-225. DOI: 10.1105/tpc.106.045047

[104] Cabral D, Banora MY, Antonino JD, Rodiuc N, Vieira P, Coelho RR, et al. The plant WEE1 kinase is involved in checkpoint control activation in nematode-induced galls. New Phytologist. 2020;**225**:430-447. DOI: 10.1111/nph.16185

[105] Haegeman A, Mantelin S, Jones JT, Gheysen G. Functional roles of effectors of plant-parasitic nematodes. Gene. 2012;**492**:19-31. DOI: 10.1016/j.gene.2011.10.040

[106] de Almeida EJ, Favery B, Engler G, Abad P. Loss of susceptibility as an alternative for nematode resistance. Current Opinion in Biotechnology. 2005;**16**:112-117. DOI: 10.1016/j.copbio.2005.01.009

[107] van Schie CCN, Takken FLW. Susceptibility genes 101: How to be a good host. Annual Review of Phytopathology. 2014;**52**:551-581. DOI: 10.1146/annurev-phyto-102313-045854

[108] Engelhardt S, Stam R, Hückelhoven R, Engelhardt S, Stam R, Hückelhoven R. Good riddance? Breaking disease susceptibility in the era of new breeding technologies. Agronomy. 2018;**8**:114-130. DOI: 10.3390/agronomy8070114

[109] Zaidi SSA, Mukhtar MS, Mansoor S. Genome editing: Targeting susceptibility genes for plant disease resistance. Trends in Biotechnology. 2018;**36**:898-906. DOI: 10.1016/J.TIBTECH.2018.04.005

Chapter 3

Biological Seed Coating Innovations for Sustainable Healthy Crop Growth in Tomato

Patta Sujatha, Madagoni Madhavi, Mandalapu Pallavi, Yarasi Bharathi, Polneni Jagan Mohan Rao, Bodduluru Rajeswari, Saddy Praveen Kumar and Anumala Akhil Reddy

Abstract

Biological seed coating (BSC) is the fastest-growing segment under the seed treatment approaches in the global seed market. It refers to the application of certain beneficial microbes to the seed prior to sowing in order to suppress, control, or repel pathogens, insects, and other pests that attack seeds, seedlings, or plants. Beneficial bioagents along with the compatible adjuvants can safely be delivered through coatings onto the seed surface. The polymer acts as a protective cover for bioagents and helps in improving the shelf life and dust-free seed. It is an efficient mechanism for placement of microbial inoculum into soil where they colonize the seedling roots and protect against soil-borne pathogens. It is also used to increase the speed and uniformity of germination, along with protection against soil-borne pathogens in nursery and improves final stand. Some induces systemic resistance in plants against biotic agents. It is a low-cost, alternative viable technology to chemical-based plant protection and nutrition. Thus, the demand for biological seed treatment solutions is increasing in view of consumer acceptance for chemical-free food. They give protection to seedlings in the nursery against damping-off fungi like *Fusarium* spp. or *Rhizoctonia* spp. and improve crop growth and yield in the main field.

Keywords: tomato, biological seed coating, innovations, seed treatment, crop growth

1. Introduction

Tomato (*Solanum lycopersicum* L.) is the second-largest global vegetable crop next to potato [1]. Tomato is known for its rich source of vitamins (A and C), minerals (potassium and folate), and antioxidants (lycopene and beta-carotene) with low calories and fat, making a healthy and balanced diet. Tomatoes are widely consumed as raw, cooked, canned, or processed products and contribute several health benefits like reduced risk of heart disease, certain cancers, and age-related macular degeneration and also improves overall digestive health. However, the production is threatened by various biotic and abiotic stresses. Among biotic stresses, tomato is affected by several fungal, bacterial, and viral diseases like damping-off (*Pythium aphanidermatum*), early blight (*Alternaria solani*), late blight (*Phytophthora infestans*), etc. Damping-off caused by *Pythium* spp. is an important nursery disease [2], which results in massive seedling death affecting qualitative and quantitative yield losses [3]. Moreover, seed and soil-borne fungi like *Phytophthora, Pythium, Rhizoctonia,* and *Fusarium* spp. can cause damping-off diseases that not only kill seeds and lower stem of seedlings but also create problems in emergence [4]. Biological stress is a complicated phenomenon that occurs in the field and is brought on by many pathogen types [5–8]. Due to non-availability of resistant cultivars in tomato, the disease is managed by the use of synthetic pesticides [3]. seed and soil-borne diseases are more difficult to control with chemical seed protectants [8–12]; chemicals used in seed treatment pose a potential hazard to human health [13–15], bees [16], birds [17, 18], and soil microbes [19, 20], which could ultimately slow down plant growth [21].

Moreover, the global biological seed treatment market size was valued at USD 1.28 billion in 2022 due to the adoption of biological seed coating by key players of seed industry in view of surging demand for chemical-free foods. It is expected to grow further at a compound annual growth rate (CAGR) of 12.4% from the year 2023 to 2030. Among these, the microbial segment contributed over 65% in 2022 and is expected to grow at a CAGR of 11.8% from 2023 to 2030. The seed protection segment is the largest market share holder among them, occupying 67%, in terms of revenue during 2022. The rising inclination of consumers toward maintaining a healthy life-style and the rising demand for chemical-free food products have fueled the increased use of biopesticides for treating vegetable seeds, accounting for the highest revenue share of over 25% in 2022 with a CAGR of 12.7% from 2023 to 2030 [22].

In view of harmful effects encountered with the usage of synthetic pesticides, there is a need to shift for alternative protection measures in vegetable production [23, 24, 25]. Several studies have proved the effectiveness of biological control agents in controlling various seed and soil-borne diseases caused by *Pythium* spp. and *Rhizoctonia solani* [26]. It has been reported that *Trichoderma* spp. was found effective against *Fusarium oxysporum and R. solani* [27], whereas the metabolites produced by *Bacillus subtilis* showed antibiotic activity in suppressing damping-off caused by *R. solani* in tomatoes. Onion seeds coated with *Pseudomonas fluorescens* F113 reduced damping-off disease [28]. In addition, the microbial agents also enhance plant growth parameters [29] by improving the uptake of micro- and macronutrients [30, 31]. For instance, *Bacillus* spp. and *Trichoderma* spp. are known for their growth promoting properties [32, 33]. In greenhouse conditions, antagonistic bacteria and fungi showed successful control of tomato diseases like blight, damping-off, bacterial and fungal wilt, powdery mildew, anthracnose, and leaf spot [34–36]. Hence, seed treatment with bioagents protects the germinating seedlings against seed and soil-borne pathogens apart from adverse abiotic conditions and thus helps in maintaining the initial plant population, which in turn will result in a 10–12% improvement in crop yield. Through the metadata analysis across crops, target pathogens, regional climates, and conditions of experiments, Lamichhane et al. [37] reported that compared to untreated seeds, the effects of biological seed treatments on seed germination, seedling emergence, plant biomass, disease control, and crop production were all considerably higher (7 ± 6%, 91 ± 5%, 53 ± 5%, 55 ± 1%, and 21 ± 2%, respectively). Moreover, biological seed coating influences germination and seedling growth by acting as biostimulants [38]. Keeping this in view, biological seed coating is an emerging technique of seed treatment in which antagonistic microbes are blended with compatible adjuvant and applied to seed surface. The bio-friendly polymer is proven to be a superior substitute for the common sugar syrup adjuvant [39, 40]. This technique takes an environmentally friendly approach by selectively using fungal antagonists to combat soil and seed-borne diseases, perhaps providing an alternative to chemical management [28]. Although the idea of treating seeds with biological agents is not new, historically, seeds are often treated prior to sowing. There are very few studies in India that advocate seed treatment with biological agents on a large scale right after cleaning and processing the seed.

2. Need of biological seed coating

The necessity for biological seed coatings is typically driven by the demand to improve crop health, reduce environmental impact, and adhere to changing regulatory standards. Biological seed coatings promote reliable and sustainable agricultural activities by utilizing the advantages of beneficial microbes and tailored/customized/adapted treatments.

The statistics on the use of biological seed coatings are meagerly available because of the diversified products and varied interests of the industry. However, due to a shift toward organic farming, and the requirement for efficient and eco-friendly seed treatments, in the recent past, a steady growth in their usage has been observed across agricultural and horticultural crops with a customized option based on crop-specific requirements. As it promotes biodiversity, BSC is a sustainable and alternative approach to conventional chemical seed treatments. The growing health concern among consumers against agrochemicals creates a wide scope for the adoption of sustainable agricultural practices and eco-friendly seed treatments for the commercialization of biological seed coatings in the near future.

Efforts are needed to improve the efficacy, stability, and compatibility of biological seed coatings, through enhancing formulations, developing improved delivery systems, and exploring new combination strains of beneficial microorganisms.

The regulations and certifications on the usage of biological seed coatings can be subjected to a particular country or region based on product efficacy, safety, and labeling. In case of organic certifications, compliance is required for specific standards with regard to biological seed treatments.

3. Benefits of biological seed coating

Through the biological seed coating technique, beneficial microbes maintain a symbiotic relationship with the seedlings and contribute to many benefits such as enhanced seed germination and seedling vigor, increased resistance toward pests and diseases, and efficient nutrient uptake and productivity.

Enhanced seed germination and seedling establishment: The microorganisms used in the seed coating can stimulate germination, improve seedling vigor, and enhance early root and shoot growth, resulting in quicker and more uniform emergence of seedlings.

Disease suppression: Certain microorganisms in the seed coating can have antagonistic effects against pathogens and suppress seed and soil-borne diseases. They may produce antibiotics or compete with harmful organisms for resources, thereby protect the seedlings from infections.

Nutrient availability and uptake: Some microorganisms, especially the mycorrhiza, has the ability to solubilize nutrients, such as phosphorus, and make them available to the developing plants. This can enhance nutrient uptake efficiency and support plant growth.

Environmental stress tolerance: Bioseed coatings contain microorganisms that produce compounds that help the plants to tolerate various environmental stresses, such as drought, salinity, or temperature extremes. These microorganisms can promote stress tolerance and improve plant survival under adverse conditions.

Reduced chemical inputs: Use of biological seed coatings, which is cost effective, can reduce the reliance on chemical treatments, such as fungicides or insecticides, as the beneficial microorganisms can provide natural protection against pathogens and pests without affecting the soil fauna.

Compatibility with other seed treatments: Bioseed coatings can be used in conjunction with other seed treatments, such as insecticides or fungicides, without significant interference. This allows for integrated pest management strategies and customized seed treatments in addition to reducing the cost of inputs.

Enhanced soil fertility: Applied BCAs continuously multiply in the soil and help in the improvement of its fertility.

4. Organisms commonly used for biological seed coating

Selection of the organism for biological seed coating depends on the crop being cultivated, benefit required, compatibility ensured, and effectiveness for specific application. The specific strains and species within the groups of organisms and their effectiveness may vary based on soil conditions, crop, and the region under consideration. It is crucial to note that the individual strains and species within these groupings of organisms might vary, and their efficacy can rely on factors including the type of crop grown, the soil, and geographical considerations. To ensure compatibility and efficacy for particular applications, the selection of the suitable organisms should be based on scientific study and field tests. Commonly used organisms for biological seed coating are included in **Table 1**.

5. Mechanism of biological control

Fungal and bacterial biological control agents use different mechanisms to control pests, pathogens, and invasive species. Specific fungal and bacterial biocontrol agents may employ multiple mechanisms simultaneously or have unique strategies depending on their characteristics and the target organisms. The selection of the appropriate biocontrol agent and mechanism depends on factors such as the target pest or pathogen, the specific crop or ecosystem, and the desired outcomes of biological control. Here are the mechanisms commonly associated with fungal and bacterial biological control:

Mutualism: It is an association between two or more species by which both species derive benefit. It contributes to biological control by fortifying the plant with nutrition and or by stimulating the host defenses. Obligatory interaction is observed between plants and mycorrhizal fungi. Facultative or opportunistic mutualism is observed between leguminous plant and *Rhizobium* bacteria.

Plant Growth Promotion: Some bacterial biocontrol agents not only suppress pests or pathogens but also promote plant growth [51]. These beneficial bacteria

Biological organism	Importance and mode of action
Trichoderma spp	Potential Biological Control Agent [41–43]. It is fast-growing, secondary opportunistic invader, having efficient rate of sporulation, which survives under adverse conditions; utilizes nutrients efficiently; produces antibiotics, cell wall-degrading enzymes like protease, beta 1–3-glucanase, and toxic metabolites [44–46]; and can induce systemic resistance in plants. *Trichoderma* species are beneficial fungi that can protect seeds and seedlings against soil-borne pathogens. After completion of the parasitic activity, host fungi cytoplasmic contents appear totally evacuated, and the hyphae look "translucent," and the hyphal "exoskeleton" remains [47]. Five isolates of *T. harzianum* were also found antagonistic to the growth of *R. solani* in dual culture on PDA with sparse to intense coiling method of myco-parasitism followed by disintegration, disorganization, and death of *R. solani* mycelium [48]. *Trichoderma* spp. can improve the plant growth and yield by enhancing the growth hormones and increment of plant beneficial microbiome [49–51].
Bacillus spp.	The endospores produced by these bacteria are resistant to adverse conditions. Due to their antagonistic properties, *Bacillus* spp inhibit plant pathogen growth through the production of antibiotics and enzymes. Examples of this group include *Bacillus subtilis, Bacillus megaterium*, and *Bacillus amyloliquefaciens*.
Pseudomonas spp	*Pseudomonas* species, such as *Pseudomonas fluorescens*, have biocontrol properties and can suppress plant diseases caused by pathogens. Through the production of antimicrobial compounds, they can suppress plant diseases by outcompeting the pathogenic organisms for resources like siderophores, iron complexes, and so forth. They can induce systemic resistance [52] and also improve seed quality and reduce bacterial canker drastically [53].
Plant Growth-Promoting *Rhizobacteria* (PGPR)	PGPR are beneficial bacteria that colonize the rhizosphere (root zone) and promote plant growth. Examples include species of *Bacillus, Pseudomonas,* and *Azospirillum*. They can improve nutrient availability, produce growth-promoting substances, and help against plant pathogens. They are used to control the bacterial spot of tomato caused by *Xanthomonas vesicatoria* [34, 54–56].
Nitrogen-Fixing Bacteria	Certain bacteria, such as species of *Rhizobium* and *Bradyrhizobium*, are used in legume crops to establish symbiotic associations in root nodules by fixing atmospheric nitrogen and making it available to the plants.
Mycorrhizal Fungi	Mycorrhizal fungi form symbiotic associations with plant roots, improving nutrient uptake, particularly phosphorus. Examples include species of *Glomus, Rhizophagus*, and so on. They explore volume of soil by extending the root system with their hyphae.
Yeasts	Some yeast species, like *Saccharomyces cerevisiae*, have biocontrol properties and can help protect seeds, seedlings, and also postharvest pathogens by releasing antifungal substances and thereby initiating the defense factors in host plant [57].
Actinomycetes	Actinomycetes are potential biocontrol agents against of soil-borne plant pathogens like *Fusarium* spp. [58], *Phytophthora* spp. [59], *Pythium* spp. [60], *Rhizoctonia* spp. [61], and *Verticillium* spp. [62].

Table 1.
Organisms used for biological seed coating (table is extrapolated from other works).

can produce plant growth-promoting substances like phytohormones, enzymes, or siderophores. They enhance nutrient uptake, improve stress tolerance, or stimulate root development, resulting in healthier plants with capacity better able to resist pest or pathogen attacks [55].

Protocooperation: It is a form of mutualism in which the involved organisms do not exclusively depend on each other for survival. Many of BCAs are facultative mutualists whose survival rarely depends on any specific host, and disease suppression depends on the prevailing environment.

Commensalism: It is a form of symbiotic interaction between organisms where one organism benefits, and the other is neither harmed nor benefited. Most plant-associated microbes are commensals with regard to host plant to which they contribute nothing, while their presence decreases the pathogen infection and disease severity.

Neutralism: Presence of one species has no effect on the other species.

Antagonism: It results in negative outcome for one or both.

Competition: The competition between and within the species results in decreased growth, activity, and fecundity. Biological control takes place when the antagonistic organism competes the pathogenic organism for nutrients in and around the host plant [55]. This will benefit beneficial organism at the expense of other. Bacterial and fungal biocontrol agents can compete with pests or pathogens for resources such as nutrients or space. They colonize ecological niches and outcompete the target organisms for nutrients and space, reducing their population size and limiting their ability to cause damage.

Parasitism: It is a type of negative symbiosis in which two unrelated organisms coexist for a long period. In this association, physically smaller parasite gets benefited by causing harm to the large organism called the host. Similarly, parasitism of virulent pathogen by beneficial organism leads to biocontrol through the stimulation of host defense systems. Some fungal biocontrol agents are mycoparasites, meaning they attack and parasitize other fungi. These biocontrol fungi invade the target pathogenic fungi, penetrate their hyphae, and extract nutrients eventually killing them. Mycoparasitic fungi can produce specialized structures like haustoria or adhesive structures to facilitate attachment and nutrient uptake [47, 48].

Hyperparasitism: Pathogen is directly attacked by the specific BCA and stops its propagation and kills it.

Hypovirulence: Reduction in disease-producing capacity of the pathogen. *Coniothyrium minitans* parasitizes sclerotia producing plant pathogens.

Predation: Killing of one organism by another for consumption and sustenance is referred as predation. For example, fungal feeding nematodes and microarthropods consume pathogen biomass.

Induction of host resistance: Some fungal biocontrol agents can induce systemic resistance (ISR) in plants [52]. When these fungi colonize plant roots or other plant tissues, they trigger the plant's defense mechanisms, leading to the production of defensive compounds or activation of systemic signaling pathways. This enhanced resistance helps plants withstand attacks from pests or pathogens. Certain bacterial biocontrol agents can trigger systemic resistance in plants similar to fungal biocontrol agents. They colonize plant surfaces or enter plant tissues, activating the plant's defense responses. This leads to the production of defense compounds, reinforcement of cell walls, or activation of signaling pathways, making the plant more resistant to pests or pathogens.

Antibiotic-mediated suppression: Many fungal biocontrol agents produce secondary metabolites, such as antibiotics, which inhibit the growth and development of target pests or pathogens [46]. These biocontrol fungi release toxic compounds into the environment that are harmful to the target organisms, suppressing their population growth or causing their death. Several bacterial biocontrol agents generate antimicrobial substances, which include antibiotics, toxins, or enzymes, which hinder the development and growth of pests or pathogens by intervening with important cellular functions, destroying cell walls, or obstructing nutrient uptake, eventually resulting in either death or inhibition of the target organisms. *Bacillus subtilis* also shown the ability to produce antibiotics and other metabolites against *Pythium* spp. [63, 64].

Suppression through lytic and other enzymes: Many BCAs generate and liberate lytic enzymes that hydrolyze polymeric compounds like chitin, proteins, cellulose, hemicellulose, and DNA and suppress plant pathogen activities directly.

6. Commercial bioformulations

These are the products available in the market that contain beneficial microorganisms or natural compounds used for various agricultural applications. Specific commercial bioformulations can vary depending on the region, crop, and regulatory approvals. These formulations are designed to enhance plant growth, improve nutrient uptake, suppress diseases and pests, and promote sustainable agriculture. The availability and formulations of these products may be subject to local regulations and market demands. Farmers and growers should consult with local agricultural suppliers or experts to identify suitable commercial bioformulations for their specific needs. Here are some examples of commercial bioformulations based on the utility:

Biopesticides: These are the formulations containing beneficial microorganisms or natural compounds used for pest management. These products may contain antagonistic bacteria, fungi, or viruses that can inhibit the growth or activity of plant pathogens. Examples include fungal biopesticides like *Trichoderma* spp., *Beauveria bassiana*, and *Metarhizium* spp. for controlling fungal diseases and targeting specific pests, *Pseudomonas fluorescens* for fungal and bacterial disease management, and *Bacillus thuringiensis* (Bt) formulations for insect control.

Biofertilizers: These formulations contain beneficial microorganisms, such as nitrogen-fixing bacteria (e.g., *Rhizobium* spp.), phosphate-solubilizing bacteria, or mycorrhizal fungi. These enhance nutrient availability and improve nutrient uptake by plants, promoting healthy crop growth and reducing the need for synthetic fertilizers.

Plant Growth Promoters: These bioformulations contain natural compounds or beneficial microorganisms that stimulate plant growth and development by enhancing root development, increase nutrient absorption, and improve stress tolerance.

Biostimulants: These are similar to PGPRs but with an added advantage of stimulating plant physiological processes and enhancing plant growth, vigor, and overall health of the plant. These products can include substances like seaweed extracts, humic acids, or plant growth-promoting rhizobacteria (PGPR).

7. Classification of commercial bioformulations

Commercial bioformulations are classified into two types based on the carrier material used 1) Solid formulation and 2) Liquid formulation.

7.1 Solid bioformualtions

Solid bioformulations refer to formulations that contain living organisms or their derivatives in a solid form. These formulations are designed for various applications, including agriculture, environmental remediation, and biotechnology. Unlike liquid formulations, which are in a liquid medium, solid bioformulations provide a different matrix or carrier for the organisms. The composition and formulation process may vary based on the specific application, target organisms, and formulation technology used. Factors like compatibility, viability, and compatibility with the carrier or matrix material need to be considered during the development and production of solid bioformulations. Solid bioformulations offer advantages like improved stability and shelf life protection for the organisms from environmental stresses, allowing for easier storage and transportation and providing controlled release of the organisms, prolonging their activity and efficacy. Solid bioformulations can be classified into several categories based on their composition and purpose.

Solid carrier based bioformulations: In this type of formulation, the living organisms are immobilized or embedded within a solid carrier material. The carrier provides physical support, protection, and nutrients for the organisms. Commercially used solid carriers are peat, vermiculite, perlite, clay, or compost. These are used for the application of biofertilizers, biocontrol agents, and bioremediation.

Pellets or Granules: These formulations are typically produced by blending the living organisms with inert solid materials and binders, and then granulating or pelletizing them into a solid form. These provide a controlled release of the organisms and allow for easier application in agriculture like biofertilizers.

Capsules or Tablets: In this type, the living organisms or their derivatives are compressed into tablet or capsule form using suitable inert materials and binders. By offering convenient handling and precise dosing, these are used in applications such as biocontrol agents or probiotics, where the organisms need to be protected during storage and transportation and delivered in a controlled manner.

Powders: Solid bioformulations can also be in the form of finely ground powders. The organisms or their derivatives are typically dried and ground into a powder, which can then be mixed with other ingredients or applied directly to the target area. Powders are commonly used for seed treatments, where the organisms are applied to seeds before planting to enhance germination; protection against pathogens, or to provide other benefits.

7.2 Liquid bioformulations

These are the formulations that contain living organisms or their derivatives available in liquid form and are designed for various applications in agriculture for managing different stresses. These formulations typically consist of a liquid medium that provides a suitable environment for the growth and survival of the organisms, along with other additives that enhance their performance. These are easy to handle, can be applied using conventional spraying equipment, and allow better distribution and colonization of the beneficial organisms on plant surfaces or in the target environment. Moreover, the liquid medium can provide nutrients and protect the organisms during storage and application. Specific liquid bioformulations vary depending on the intended application, target organisms, and the specific formulation technology employed. The formulation process may involve selecting appropriate strains, optimizing growth conditions, and adding stabilizers or additives to enhance shelf life and efficacy. Liquid bioformulations can be broadly categorized into two types: microbial bioformulations and biopesticides.

Microbial Bioformulations: These formulations contain beneficial microorganisms, such as bacteria, fungi, or algae, which are used for various purposes. Some common examples include, Biofertilizers: These formulations contain nitrogen-fixing bacteria or other beneficial microorganisms that enhance soil fertility and plant nutrition. They can improve nutrient availability, promote plant growth, and enhance crop yield. Biocontrol agents: These formulations consist of beneficial microorganisms that help control plant diseases and pests. They can suppress the growth of pathogenic microorganisms or insects, providing a natural and environmentally friendly alternative to chemical pesticides. Biostimulants: These formulations contain microorganisms or their metabolites that stimulate plant growth, enhance nutrient uptake, or improve stress tolerance. They can be used to enhance crop productivity and improve plant health. Bioremediation agents: These formulations contain microorganisms capable of degrading or detoxifying pollutants in the environment. They are used for the cleanup of contaminated soils, water bodies, and industrial sites.

Biopesticides: These formulations are specifically designed for pest control and contain natural substances derived from living organisms. They can be based on microorganisms, such as bacteria, viruses, or fungi, or the plant extracts. Biopesticides are considered safer and more environmentally friendly compared to synthetic chemical pesticides.

8. Methods of biological seed coating

Biological seed coating is a technique used to enhance seed performance and protect seeds from various stresses, such as pathogens, pests, and adverse environmental conditions. There are several methods of biological seed coating that can be employed, each with its advantages and applications. The specificity of the method and formulation varies based on the crop, target pests or pathogens, desired outcomes, and the prevailing environmental conditions. These biological seed coatings are not mutually exclusive, but combinations of different coatings can be used to achieve multiple benefits. Biological seed coatings can be classified into several types based on their composition, purpose, and the agents used. Here are some common types of biological seed coatings.

Encapsulation: This method involves encapsulating the seeds with a protective layer made of natural materials, such as clays, polymers, or biodegradable substances. The coating layer acts as a barrier against external factors, such as pathogens and insects, while providing a controlled release of nutrients or bioactive compounds.

Microbial Seed Coating: It involves application of beneficial microorganisms to the seed surface. These microorganisms can include beneficial bacteria, fungi, or mycorrhizal fungi. The coating provides a reservoir of beneficial microbes that can enhance seed germination, nutrient uptake, and overall plant growth. Additionally, these microbes can help suppress the growth of harmful pathogens.

Biopolymer Coating: Biopolymer-based coatings are derived from natural polymers, such as chitosan, alginate, or starch. These coatings can provide protection against pathogens, regulate water absorption, and enhance seed adhesion. Biopolymer coatings are often used to improve seed viability, reduce seed-borne diseases, and promote seedling establishment.

Biopesticide Coating: Biopesticide seed coatings involve applying naturally derived plant extracts or biocontrol agents to the seed surface. These coatings can help protect seeds from pests and diseases, such as insects, nematodes, or fungi. Biopesticide coatings provide an eco-friendly alternative to chemical pesticides while maintaining seed quality and promoting healthy plant growth.

Nutrient Coating: Nutrient seed coatings aim to provide essential nutrients to the germinating seed or seedling. These coatings contain fertilizers, growth-promoting substances, or micronutrients. The coating ensures that the seed has access to necessary nutrients during early growth stages, enhancing germination, early vigor, and establishment.

Hormone-based Coating: The method involves application of plant growth regulators, such as auxins, cytokinins, or gibberellins, to the seed surface. These hormones can influence seed germination, root development, and overall plant growth. Hormone-based coatings are used to enhance seedling vigor, stimulate root growth, and improve stress tolerance.

9. Factors that affect the effectiveness of biological seed coating

The following factors influence the effectiveness of the biological seed coating.

Seed characteristics: The size, shape, and texture of seeds may vary depending on the type of seed. These traits may affect the efficacy of biological coatings adherence and persist on seed surface. Some seeds may have natural structures or coatings that make it easier or hinder the microbes to attach.

Choice of Microorganism: The choice of selecting microorganisms for seed coating is crucial. Factors like microbial strains, its compatibility with specific crop, and the expected benefits (e.g., disease/pest management, nutrient solubilization) should be taken into consideration as varied microbes have their own requirements for growth, survival, and interaction with seed and seedlings.

Colony-Forming Units per milliliter (CFU/ml) or per gram (CFU/g): It allows the quantification of viable microbes and assessment of microbial load, growth, or inhibition in different samples and systems which is important for disease suppression.

Coating formulation: To ensure viability and stability, the composition of the microorganisms along with their formulation is important. The survival and efficacy of these microbes can be affected by the factors such as carrier materials, protective additives, and application techniques. However, optimization of coating formulation is required to provide ambient conditions for the growth and activity of the microbes.

Application technique: The process used for applying biological coatings to seeds has a direct effect on the coating's coverage, consistency, and adherence. The techniques viz., film coating, slurry coating, or vacuum impregnation can be employed.

Adjuvant used for coating: Bio-friendly polymer and nonionic polymers, biopolymers used as an adjuvant recorded good viability and shelf life of bioagent compared to conventional adjuvants like sugar/jaggery syrup.

Viability and shelf life: The ability of a bioagent to survive and maintain its effectiveness over time when stored under particular conditions is referred to as its viability and shelf life. The bioagent must be viable and have a long shelf life in order to be useful across a range of applications.

Environmental conditions: Environmental factors, including temperature, humidity, and light exposure, can influence the survival and activity of the microorganisms on the seed surface. Some microorganisms may have specific temperature or moisture requirements for optimal performance. High temperatures or prolonged exposure to UV light can negatively impact the viability and efficacy of the microorganisms.

Seed storage and handling: Proper seed storage and handling practices are essential to maintain the viability of the biological coating. Storage conditions, such as temperature, moisture, and duration, can affect the survival of the microorganisms. It is important to follow recommended storage guidelines to preserve the efficacy of the biological coating until the seeds are planted.

Interactions with other seed treatments: If other seed treatments, such as fungicides or insecticides, are applied in conjunction with biological seed coating, compatibility and potential interactions should be considered. Some chemical treatments may adversely affect the survival or activity of the microorganisms, leading to reduced efficacy of the biological coating. Considering these factors, the seed coating process can be optimized accordingly, and the effectiveness can be maximized, leading to the improvement of seed quality and crop performance.

10. Several factors can influence the viability and shelf life of a bioagent

It is significant to remember that a bioagent's viability and shelf life might change based on the particular organism, formulation, and storage conditions. Manufacturers often include guidelines and suggestions for storage conditions, shelf life, and handling procedures. Following these recommendations will assist the bioagent to remain as viable and effective as possible throughout its shelf life.

Storage Conditions: The viability and shelf life are greatly influenced by the storage conditions, which include temperature, humidity, and light exposure. The storage conditions for each bioagent should be specified based on their specific requirements. For instance, some bioagents can be stored at ambient temperature, while others may need to be refrigerated. It is crucial to adhere to the manufacturer's prescribed storage recommendations.

Formulation and Packaging: The formulation and packaging of the bioagent can impact its viability and shelf life, wherein the selection of carrier materials, stabilizers, and protective additives can help in enhancing the stability and longevity of the bioagent. Further, an airtight and moisture-resistant container prevents contamination and moisture exchange, which may affect the viability of the bioagent.

Strain Selection: It is important to select stable strains or isolates of a bioagent as some of them may have better survival and stability potential and more suitable for extended shelf life. Screening and selection of strains help in identifying the strains with preferred characteristics.

Adjuvant used for coating: Bio-friendly polymer not only acts as a good binder but also provides nutrients required for the survival of microorganisms on the coated seed surface compared to sugar/jaggery syrup.

Quality Control: In order to ensure bioagent viability and shelf life, it is much essential to follow strict quality control measures during production and formulation, which helps in assessment of viability and enumerating the microbes to ensure consistent quality of a product. Further, batch testing and monitoring at regular intervals are needed to identify any potential deviations or reduction in viability.

Shelf Life Testing: Stability test should be conducted to evaluate the viability and efficacy of the bioagent over a period of time by subjecting the bioagent to accelerate aging or ambient storage conditions and assessing its viability and efficacy at periodic intervals. The test helps in determining the expiry period of the product.

11. Compatibility studies for biological seed coatings

Studies on the compatibility between the biological agents used in the formulation of biological seed coatings and the other components of the coating system have to be performed to ensure that the biological agents remain effective and efficient and do not adversely affect the coating or other additives. Key components to consider when executing compatibility studies for biological seed coatings are:

Compatibility with Coating Materials: Biological agents have to be tested for their compatibility with coating materials, such as binders, adhesives, polymers, film-forming agents, and so on, where they do not interfere with the coating process, adhesion to the seed surface, as well as physical integrity of the coating. Other additives like colorants, surfactants, or nutrients that are included in the seed coating formulations should be evaluated for their compatibility with biological agents, where the additives do not have negative impact on viability or efficacy of biological agents. These studies can be carried out by following poison food technique for fungi and zone of inhibition technique for bacteria. In agricultural or horticultural crops, to ensure the successful integration of the biological agents into the coating formulation, the compatibility studies help in the optimization of the formulation and the performance of coated seeds to confer desired benefits.

Compatibility with other bioagents: For the development of bioagents consortia, biological agents have to be tested for their compatibility with each other with the help of dual culture technique.

Viability Assessment: The effect of the coating materials on the viability of the biological agents has to be assessed by determining the Colony Forming Units (CFU) to assess the survival and growth of microbe after subjecting to the coating process.

Physical and Chemical Stability: The biological agents have to be assessed for their physical as well as chemical stability to observe any deviations in pH, temperature, or moisture content, which affect the viability of microbes. Performing long-term stability studies influences the shelf life and storage conditions of coated seeds.

Efficacy Evaluation: Finally, after seed coating, the effectiveness of biological agents and further the ability of the coated seeds to have desired biological benefits like improved germination, plant growth promotion, disease or pest control should be assessed. The efficacy of bioagents can be tested both in the laboratory and field conditions.

12. Materials required for biological seed coating

The specific material requirements will be varied and depend on target crop, bioagent strains, coating formulation, and desired outcomes. For effective application of biological agents onto the seed surface, the following are the commonly used materials in biological seed coating:

Beneficial Microorganisms: Selection of potential strains of beneficial microorganisms (bacteria, fungi, or mycorrhizal fungi) in pure form or commercial products from specialized suppliers. The use of effective native isolates will give better protection against disease causing pathogens.

Carrier Materials: Carrier Materials: Talc is a commonly used carrier because of its fine texture, absorption capacity, and ability to aid in the equal dispersion of microorganisms on the seeds. This helps to offer a suitable medium for the beneficial microbes to adhere to the seed surface. Clays, like kaolin or bentonite, offer the microorganisms a moderate amount of adhesion as well as protection. Vermiculite is a type of mineral that can hold moisture and create an ideal habitat for the development of microorganisms. Biopolymers: Chitosan, alginate, or starch are examples of biodegradable polymers that can be utilised to enhance adhesion, microorganism protection, and controlled nutrient release.

Adhesive Substances/Adjuvants: Plant-based gums, gelatinous solutions, or biodegradable polymers like starch or cellulose are used to increase the adherence of the carrier and microorganisms to the seed surface. These increase stickiness and encourage the adhesion of the carrier material to the seed.

Protective Agents: To enhance seed protection, protective agents may be added in the coating formulation, such as antioxidants like ascorbic acid or vitamin E, to provide protection to the microorganisms and seeds from oxidative stress during storage or handling. Natural or synthetic compounds having antifungal or antibacterial properties can be included to the coating formulation to protect against seed-borne pathogens.

Proper packaging materials (moisture-proof) and packing method is needed to maintain seed viability and integrity of the coating.

13. Dosages of biological seed coating

Although dosages for biological seed coating vary depending on the product, crop, seed size, and so forth, it is necessary to account for the amount and active ingredients applied to the seed. Therefore, to determine the accurate dosages for any particular application, valuable suggestions have to be taken from product manufacturer, agricultural extension services, or agronomists. The following are some points to be considered for fixing dosages of biological seed coating:

Recommended Application Rates: For the particular product, manufacturers frequently offer recommended application rates or instructions. Depending on the target crop, the type of seed, and the desired results, these guidelines may recommend a range of treatment rates. To ensure proper dose, it is essential to carefully read and adhere to these recommendations.

Seed Size and Weight: The dosage when coated on the seed can influence the size and weight of the seeds. Generally lower dosage is required for the smaller seeds over the larger seeds to achieve maximum coverage and adherence of the coating. Uniform application of seed coatings should be done to have consistent results.

Concentration of Active Ingredients: The dosage can also be affected by the concentration of active components, such as beneficial microbes or bioactive substances, in the composition of the seed coating. The suggested dosage will depend on the exact formulation because various products may have variable concentrations.

Seed Coating Method: The dosage of application may vary depending on the method by which the seeds are coated. In order to ensure sufficient coverage and adherence of the coating to the seed surface, the dosage may need to be adjusted for various coating processes, such as slurry coating, dry coating, or film coating.

Seed Treatment Equipment: The dose of the biological seed coating can vary depending on the type of seed treatment equipment being utilized. In order to provide precise and constant application rates, manufacturers may offer guidelines for equipment calibration and settings.

14. Process of biological seed coating

In order to ensure that the biological agents are applied accurately and that the seeds are properly treated, the biological seed coating process entails several steps. It is important to keep in consideration that the precise procedures and methods may change based on the biological agents, seed coating components, and equipment utilized. Additionally, the seed coating procedure may be influenced by factors including seed species, desired results, and regulatory considerations. The common procedure includes:

Selection of Biological agents: Select potential beneficial microbes such as bacteria, fungi, mycorrhizal fungi, or other PGPRs for seed coating that offer protection against pathogens or enhance nutrient uptake.

Preparation of Biological agents: Culture the bioagents in a suitable growth medium under controlled conditions to have a sufficient spore load/population for seed coating.

Seed Treatment: Prior to treatment, the seeds should be cleaned, sorted, and evaluated for quality. Remove any contaminated or unhealthy seeds. If necessary, treatments like applying fungicides or insecticides can be done at this stage.

Coating Formulation: By combining the biological agents with a carrier substance, prepare the coating formulation. The carrier material might be an adhesive substance like biopolymers or gelatinous solutions, or it can be a powdered material like talc, clay, or vermiculite. The biological agents should adhere to the seed surface, and formulation should be distributed uniformly.

Seed Coating Application: Apply the coating formulation on to the seeds, which can be done either manually or with specialized seed coating tools. In order to obtain a consistent coating, the seeds are often tumbled or rotated in a drum, while the coating composition is sprayed or applied evenly.

Drying and Curing: After application of coating, to achieve efficient coating adhesion and to avoid seed clumping or sticking, the coated seeds must be dried and cured. This can be accomplished by distributing the coated seeds in a well-ventilated area or by employing temperature and humidity-controlled drying chambers.

Quality Control and Packaging: Following drying and curing, the coated seeds are subjected to quality control inspections to make sure the coating application is uniform and meets the standards. After being packaged in appropriate containers with the correct labeling and storage conditions, the coated seeds are stored to maintain seed viability advanced innovations in biological seed coating for tomato are mentioned in **Table 2**.

15. Mode of action of biological seed coating

Biological seed coatings have different modes of action. Depending on the microorganisms employed and their interaction with the seed and the surrounding environment, the precise method of action can change. Here are a few typical biological seed-coating modes of action:

Biological Control: The beneficial microbes used in seed coatings show antagonistic properties and combat the plant pathogens through various mechanisms like antibiosis [47], competition [48], and parasitism, apart from inducing systemic resistance in host plants [52, 67, 75, 86, 95].

Disease Suppression: Many of the microbes used in seed coating technique protect the seeds as well as seedlings from various seed and soil-borne infections by producing antibiotics or other antimicrobial compounds that affect the growth of pathogens [65]. In addition to competing for resources, they colonise the rhizosphere, which causes host plants to develop systemic resistance and so protect themselves from plant infections [73, 79, 94].

Nutrient Solubilization and Enhancement: Some microorganisms have the capacity to solubilize nutrients. Nutrients like phosphorus can be solubilized by specific microbes, which increase their availability to plants [93]. They synthesize the phosphatases and other enzymes that convert complex nutrient forms into simpler forms that plants can readily absorb. This promotes plant growth and development and improves the efficiency of nutrient uptake.

Enhanced Germination and Seedling Establishment: Microorganisms found in seed coatings can promote seed germination and increase the vigor of seedlings [39, 68]. These microbes may produce enzymes that stimulate root and shoot growth or break down seed dormancy. As a result, seedlings emerge more quickly and uniformly, enhancing the crop establishment [81, 89, 90, 92].

Plant Growth Promotion: Beneficial microbes used in seed coatings enhance plant growth-promoting substances like phytohormones, which can directly encourage the plant growth (65, 71, 74 82). However, they indirectly aid in increasing nutrient availability, promoting root development, and enhancing physiological plant processes leading to health plant growth [94, 98].

Environmental Stress Tolerance: Some microorganisms found in seed coatings can increase plants' resistance to environmental challenges like drought, salt, and severe temperatures [70, 71, 83]. They could synthesize stress-responsive compounds, such as osmoprotectants or heat-shock proteins, which aid plants in surviving adverse conditions and increasing their yield [85, 88].

16. Precautions for biological seed coating

Certain safety guidelines should be followed while utilizing biological seed coatings to ensure correct handling, application, and security. However, these precautions are only general recommendations, and they may change depending on the product, formulation, and regional regulations. Therefore, always read the product label and consult the manufacturer or industry professionals regarding specific usage guidelines and safety measures before using of the biological seed coating product. When handling biological seed coatings, the following general safety measures have to be taken:

Follow Product Instructions: Read the manufacturer's instructions, suggestions, and safety data sheets that are provided with the biological seed coating product carefully, and then follow them. Follow the manufacturer's recommended dosage, application guidelines, and safety instructions.

Personal Protective Equipment (PPE): As suggested by the product's manufacturer, wear proper protective equipment, such as gloves, protective clothes, goggles,

Crop	Beneficial organism	Method	Mode of action	Reference
Tomato	*Trichoderma harzianum*	Seed treatment	*Trichoderma hyphae completely engulfed the hyphae of R. solani by coiling and hooking.* Decreased 49.5 and 64.33% of pre-emergence damping-off disease caused by *P. infestans* and *R. solani*, respectively. *Trichoderma* treatments had a positive effect on growth parameters compared to control.	[65]
Tomato	Five biocontrol bacteria (*Bacillus amyloliquefaciens* (Ba), *Bacillus subtilis* (Bs wy-1), *Bacillus subtilis* (WXCDD105), *Pseudomonas fluorescens* (WXCDD51), and *Bacillus velezensis* (WZ-37)	Mixed with auxiliary factors (inactive components of seed-coating agent) after fermentation	The seedling mortality rate due to *Pythium aphanidermatum* was 26.7% lower than that of the sterile water control and 20% lower than that of carbendazim. The seedling mortality rate caused by *Fusarium* spp. was 44.31% lower than that of the control and 22.36% lower than that of carbendazim.	[66]
Tomato	*Bacillus siamensis* CU-XJ-9	—	Produced nodules and destroyed the mycelial structure of *Fusarium graminearum* through the production of lipopeptide antibiotics.	[67]
Pigeonpea	*Trichoderma viride, Pseudomonas fluorescens, Bacillus subtilis*	Seed Coating	Improved seed germination in biologically coated seed	[68]
Tomato	Endophytic bacteria (SuRW02)	Seed coating	Endophytic bacteria recorded 38% incidence of Fusarium wilt and disease severity indices of 0.37. The uncoated seeds and seed coating without endophytic bacteria showed disease incidences of 70 and 50% and disease severity indices of 2.00 and 1.25, respectively. It further promoted tomato plant growth and quality of tomato production.	[69]
Tomato	*Trichoderma pseudokoningii*	Seed biopriming with vermiwash combination	Under heat-stress conditions, root biomass increased.	[70]
Tomato	*Bacillus subtilis subsp. subtilis,* and *T. harzianum*	—	Significantly enhanced tomato plant growth and immunity when used against *P. infestans.*	[71]

Crop	Beneficial organism	Method	Mode of action	Reference
Tomato	Four *Trichoderma* isolates	Seed treatment	Depending on the *Trichoderma* isolate, there was a significant rise in germination percentage and a reduction in the incidence of pre-emergence damping-off with biocontrol efficiency against *R. solani* ranging from 20.66 to 39.23% and *Pythium* spp. from 32.39 to 64.46%. However, isolate *T. harzianum* has improved plant height, number of leaves and flowers per plant, dry and fresh weight, root length and yield	[72]
Chili	*T. harzianum*	Seed treatment	Proved effective in controlling damping-off disease caused by *Pythium aphanidermatum*	[73]
Tomato	*T. harzianum*	Seed treatment	Overall enhancement of plant growth was observed when used against *Pythium ultimum* and *Phytophthora capsica*.	[74]
Pigeonpea	*Trichoderma viride, Pseudomonas fluorescens, Bacillus subtilis, Rhizobium* spp.	Seed treatment	Improved seed germination in biologically coated pigeonpea	[39]
Tomato	*T. harzianum*	Seed treatment	Isolate Th-Sks showed 79.47% growth inhibition of *Phytophthora infestans* with the suppression efficacy of 91–100% in field and promoted plant height and fruit yield	[75]
Tomato	Isolates of *T. asperellum* TRC 900 (10^6 spores/ml) and *B. subtilis* BS 01 (10^6 CFU/ml)	Three grams of wet seeds were mixed either with *B. subtilis* or with *T. asperellum* suspensions followed by air drying at 25°C for 24 h.	Under heavy disease pressure, coated seed showed lower percentage of pre-emergence damping-off. Combination of seed coating and fertilizer application (NPK fertilizer of 400 ppm) accelerated the growth of the seedlings. Among the two bioagents, *B. subtilis* was highly efficient in controlling damping-off caused by *P. aphanidermatum* compared to *T. asperellum*.	[76]
Tomato	*Trichoderma* spp.	Different substrates amended with *Trichoderma*	Noticeable reduction of damping-off caused by *R. solani* in tomato was seen.	[77]

Crop	Beneficial organism	Method	Mode of action	Reference
Tomato	*T. harzianum*	Foliar application of *T. harzianum* spores	Inhibited 67.78% incidence of *Alternaria* leaf blight	[78]
Tomato	Endophytic actinomycetes (Strains CA-2 and AA-2 related to *Streptomyces mutabilis* NBRC 12800 T and *Streptomyces cyaneofuscatus* JCM 4364 T)	Seed coating	Reduced the severity of damping-off of tomato seedlings and showed a substantial rise in the seedling fresh weight, seedling length and root length of the treated seedlings compared to the control.	[79]
Tomato	*Bacillus* spp. and *Pseudomonas* spp. (PGPR)	Seed coating	Inhibited the growth of the *Fusarium* spp., a wilt pathogen, by making use of mechanisms such as indole acetic acid production, siderophore production, phosphate solublilization, systemic resistance induction and production of antifungal volatile componuds.	[80]
Tomato and chili seedlings	Antagonistic *Streptomyces rubrolavendulae* S4	Growing of seedlings in *P. infestans* artificially inoculated peat moss	Significant increase in the survival rates of colonized tomato and chili seedlings, from 51.42 to 88.57% and 34.10 to 76.71%, respectively.	[81]
Tomato	Lactic acid bacteria (LAB), isolated from milk and yoghurt	Seed treatment or soil drench	Acted as plant growth promoting bacteria and biocontrol agent against some phytopathogenic fungi such as *Fusarium oxysporum*, under *in vivo* tests	[82]
Tomato	*Bacillus subtilis* (GIBC-Jamog) and *Burkholderia cepacia* (TEPF-Sungal) and PGPR strain mixtures, S2BC-1 (*B. subtilis*) + GIBC-Jamog (*B. subtilis*) and S2BC-2 (*Bacillus atrophaeus*) + TEPF-Sungal (*Burkholderia cepacia*)	Seed bacterization and soil application of S2BC-1 + GIBC-Jamog challenge-inoculated with *F. oxysporum* f.sp. *lycopersici*.	Significant decrease in the incidence of vascular wilt caused by the fungus *Fusarium oxysporum* f.sp. *lycopersici* and localized induced systemic resistance (ISR) compared to non-bacterized seed. .	[83]
Chili, Tomato and brinjal	*P. fluorescens*	Biopriming	Stimulated the germination of seeds better than some fungal biopriming agents *viz.*, *T. viride* AN-10 and *T. harzianum* AN-13	[84]

Crop	Beneficial organism	Method	Mode of action	Reference
Tomato	*T. harzianum* Rifai strain T-22	—	Alleviated abiotic stress tolerance factors like osmosis, salinity, chilling, and high temperature.	[85]
Tomato	*B. brevis*	—	Potential biological control agent that reduced the impact of *Fusarium oxysporum* f.sp. *lycopersici*	[86]
Tomato	*T. harzianum* and fluorescent Pseudomonas	seed bio-priming	Increased seed germination (22–48%), decreased germination time (2.0–2.5 days), and decreased incidence of wilt in pots and fields. In pots and the field, respectively, the combination of fluorescent Pseudomonas, *T. harzianum,* and arbuscular mycorrhizal fungus (AMF) recorded superior control than uninoculated treatment. The yield was also increased by 20% by the combination treatments. In all treatments, the addition of cow dung compost (CDC) further decreased the disease and improved yield in all treatments.	[87]
Tomato	*Trichoderma* spp.	Seed treatment	In the presence of abiotic stress oxidative damage conditions, *Trichoderma* treatment guarantees a high speed and more uniform germination through physiological protection and decreased accumulation of lipid peroxides.	[88]
Tomato	*T. harzianum* and *P. fluorescens*	Seed coating of inoculum	Reduced the mean germination time to less than 2.5 days and increased the germination rate to more than 48%. In comparison to single-isolates, inoculant combinations were more successful.	[89]
Carrot and onion	*T. viride, Pseudomonas fluorescens, Pseudomonas chlororaphis, Clonostachys rosea,* and *Pseudomonas chlororaphis*	Seed priming	Improved emergence and a better emergence time under glasshouse condition.	[90]

Crop	Beneficial organism	Method	Mode of action	Reference
Tomato	*T. harzianum*	Seed coating	Recorded high seedling emergence and seedling shoot fresh weight.	[91]
Tomato	*Pseudomonas fluorescens (10^8 cfu/ ml)*	Biological seed treatment- Slurry seed treatment at the rate of 10 g/ kg seeds followed by air drying for 12 h.	Improved seed quality and significantly reduced *Xanthomonas vesicatoria*causing bacterial spot disease incidence in fields	[92]
Tomato	*T. harzianum*	Plants treatment	Showed reduction in damping-off and root rot diseases. Gave high fruit yields.	[93, 94]
Tomato	*Trichoderma* spp.	Seed treatment	Considerable yield increase was noticed in the plant seeds when pre-treated with *Trichoderma* spore suspension.	[35, 94]
Tomato	*Trichoderma* spp.	Biopriming of seeds with *Trichoderma* formulations before sowing/planting	The ability of isolates of *T. harzianum* to produce phytohormones such auxins, gibberellins, and cytokinins, vitamins, and solubilizing minerals contributed to the promotion of tomato growth parameters besides, their role in direct inhibition of pathogen growth.	[93]
Tomato	*B. subtilis* and *B. lentimorbus*	—	Potential to be used as biocontrol agents against *R. solani*	[95]
Tomato	*Pseudomonas aureofaciens* AB254 (10^5 bacteria cfu / seed.)	Seeds were bio-osmoprimed by soaking in aerated −0.8 MPa $NaNO_3$ for 4 days at which time a mixture of nutrient broth, polyalkylene glycol, and bacterial stock were added. Seeds were then hydrated for an additional 3 days	At a slightly slower rate, bio-osmopriming also offered defense against the damping-off fungi *Pythium ultimum*. This method also increases the possibility of the seed lot establishing a healthy stand.	[96]
Tomato	*Pseudomonas aureofaciens* AB254 (10^8 bacteria cfu / seed.)	Seed coating with AB254	AB254 coatings protected tomato seeds from damping-off fungi *Pythium ultimum* infection equally as the fungicide, Metalaxyl.	[96, 97]

Crop	Beneficial organism	Method	Mode of action	Reference
Tomato	*T. harzianum*	Seed coating or in wheat-bran/peat (1:1,v/v) rooting mixture	*Fusarium oxysporum* f.sp. *radicis-lycopersici* was completely eradicated from the rhizosphere root zone due to the proliferation of *T. harzianum* and resulted in 26.2% increase in tomato yield over the control.	[98]

Table 2.
Advanced innovations in biological seed coating in tomato (table is extrapolated from other works).

and respiratory protection. Refer to the product label for detailed instructions since PPE requirements can vary depending on the particular product and application method.

Storage and Handling: The biological seed coating materials should be kept out of direct sunlight and high temperatures in a cool, dry location. Observe any special storage instructions that the manufacturer may have provided. Be cautious when handling the items to prevent spills, leaks, and contact with the skin, eyes, or clothing.

Mixing and Application: The biological seed coating product should be prepared and mixed in accordance with the manufacturer's instructions. Use the right equipment and adhere to the suggested application techniques. To reduce the risk of exposure or inhalation, avoid creating too much dust or aerosol during mixing and application.

Environmental Considerations: When applying biological seed coatings, adhere to the appropriate environmental rules and regulations. Applying the coatings in areas with possible environmental hazards, such as those close to water bodies or delicate habitats, is not advised. Precautions should be taken to prevent water sources from being contaminated, including adhering to buffer zone restrictions.

Seed Quality: Make sure that the seeds being coated are of high quality and not contaminated by insect or pathogens, or other pollutants. Poor quality seeds may increase risks or have an adverse influence on the seed coating's efficiency.

Disposal: Using correct waste disposal techniques to dispose of any empty containers, unused product, or waste materials in accordance with local regulations taking to consideration their impact on the environment.

Record Keeping: Ensure that all information regarding the products used, application rates, dates, and any observations or results is accurate. This knowledge may be useful for future research, assessment, or troubleshooting.

17. Future challenges

Manipulation or improvement of the PGPR strains for overall health of the crop will determine their future. However, genetically engineering of the PGPR to obtain desired effects and employing nano-fertilizers or/and pesticides combined with isolated effector molecules from the PGPRs instead of the complete organism are the few approaches that can be used to accomplish the aspects. Further, the technology or/and the technological improvements in this area have yet to be demonstrated, assessed, and standardized, and they should also be economically viable.

18. Conclusion

It is to conclude that the biological seed coating offers environmentally friendly and sustainable approach that provides efficient protection against seed and soil-borne pathogens, seed quality improvement, and crop productivity enhancement by reducing reliance on harmful synthetic inputs of agriculture. It is in line with the principles of sustainable farming practices, promotes ecological balance, and supports the safe and healthy food production for a growing population.

Author details

Patta Sujatha*, Madagoni Madhavi, Mandalapu Pallavi, Yarasi Bharathi, Polneni Jagan Mohan Rao, Bodduluru Rajeswari, Saddy Praveen Kumar and Anumala Akhil Reddy
Department of Seed Science and Technology, Seed Research and Technology Centre, Professor Jayashankar Telangana State Agricultural University, Hyderabad, India

*Address all correspondence to: patta.sujatha@gmail.com

References

[1] F.A.O. 2022. FAOSTAT. Available from: www.faostat.fao.org.

[2] Babalola OO, Glick BR. Indigenous African agriculture and plant associated microbes: Current practice and future transgenic prospects. Scientific Research and Essays. 2012;7:2431-2439

[3] Muriungi SJ, Mutitu EW, Muthomi JW. Efficacy of cultural methods in the control of *Rhizoctonia solani* strains causing tomato damping-off in Kenya. African Journal of Food, Agriculture, Nutrition and Development. 2014;**14**:124-147

[4] Rehman SU, Lawrence R, Kumar EJ, Badri ZA. Comparative efficacy of *Trichoderma viride, T. harzianum* and carbendazim against damping-off disease of cauliflower caused by *Rhizoctonia solani* Kuhn. Journal of Biopesticides. 2012;**5**(1):23-27

[5] Lamichhane JR, Venturi V. Synergisms between microbial pathogens in plant disease complexes: A growing trend. Frontiers in Plant Science. 2015;**6**:385. DOI: 10.3389/fpls.2015.00385

[6] Rojas JA, Jacobs JL, Napieralski S, et al. Oomycete species associated with soybean seedlings in North America—Part II: Diversity and ecology in relation to environmental and edaphic factors. Phytopathology. 2016;**107**:293-304. DOI: 10.1094/PHYTO-04-16-0176-R

[7] van Agtmaal M, Straathof A, Termorshuizen A, et al. Exploring the reservoir of potential fungal plant pathogens in agricultural soil. Applied Soil Ecology. 2017;**121**:152-160. DOI: 10.1016/j.apsoil.2017.09.032

[8] You MP, Lamichhane JR, Aubertot J-N, Barbetti MJ. Understanding why effective fungicides against individual soilborne pathogens are ineffective with soilborne pathogen complexes. Plant Disease. 2020;**104**:904-920. DOI: 10.1094/pdis-06-19-1252-re

[9] Rossman DR, Byrne AM, Chilvers MI. Profitability and efficacy of soybean seed treatment in Michigan. Crop Protection. 2018;**114**:44-52. DOI: 10.1016/j.cropro.2018.08.003

[10] Mourtzinis S, Krupke CH, Esker PD, et al. Neonicotinoid seed treatments of soybean provide negligible benefits to US farmers. Scientific Reports. 2019;**9**:11207. DOI: 10.1038/s41598-019-47442-8

[11] Fadel Sartori F, Floriano Pimpinato R, Tornisielo VL, et al. Soybean seed treatment: How do fungicides translocate in plants? Pest Management Science. 2020;**76**:2355-2359. DOI: 10.1002/ps.5771

[12] Lundin O, Malsher G, Högfeldt C, Bommarco R. Pest management and yield in spring oilseed rape without neonicotinoid seed treatments. Crop Protection. 2020;**137**:105261. DOI: 10.1016/j.cropro.2020.105261

[13] White KE, Hoppin JA. Seed treatment and its implication for fungicide exposure assessment. Journal of Exposure Analysis and Environmental Epidemiology. 2004;**14**:195-203. DOI: 10.1038/sj.jea.7500312

[14] Han R, Wu Z, Huang Z, et al. Tracking pesticide exposure to operating workers for risk assessment in seed coating with tebuconazole and carbofuran. Pest Management Science. 2021;**77**(6):2820-2825. DOI: 10.1002/ps.6315

[15] AGRICAN. Enquête agriculture & cancer. 2020. p. 60. Available from: www.agrican.fr

[16] Rundlof M, Andersson GKS, Bommarco R, et al. Seed coating with a neonicotinoid insecticide negatively affects wild bees. Nature. 2015;**521**:77-80. DOI: 10.1038/nature14420

[17] Li Y, Miao R, Khanna M. Neonicotinoids and decline in bird biodiversity in the United States. Nature Sustainability. 2020b;**3**:1027-1035. DOI: 10.1038/s41893-020-0582-x

[18] Fernández-Vizcaíno E, Ortiz-Santaliestra ME, Fernández-Tizón M, et al. Bird exposure to fungicides through the consumption of treated seeds: A study of wild red-legged partridges in Central Spain. Environmental Pollution. 2021;**292**:118335. DOI: 10.1016/j.envpol.2021.118335

[19] Nettles R, Watkins J, Ricks K, et al. Influence of pesticide seed treatments on rhizosphere fungal and bacterial communities and leaf fungal endophyte communities in maize and soybean. Applied Soil Ecology. 2016;**102**:61-69. DOI: 10.1016/j.apsoil.2016.02.008

[20] Gomes A, Mariano RL, Silveira EB, Mesquita JC. Isolation, selection of bacteria, and effect of *Bacillus* spp. in the production of organic lettuce seedlings. Horticultura Brasileira. 2003;**21**:699-703

[21] Vasanthakumari MM, Shridhar J, Madhura RJ, et al. Role of endophytes in early seedling growth of plants: A test using systemic fungicide seed treatment. Plant Physiology Reports. 2019;**24**:86-95. DOI: 10.1007/s40502-018-0404-6

[22] Available from: http://www.fortunebusinessinsights.com/vegetable-seed-market-103066

[23] van den Boogert H, AJG L. Compatible biological and chemical control systems for *Rhizoctonia solani* in potato. European Journal of Plant Pathology. 2004;**2**:110. DOI: 10.1023/B:EJPP.0000015325.33299.e0

[24] Rauf C, Ashraf M, Ahmad I. Management of black scurf disease of potato. Pakistan Journal of Botany. 2007;**39**(4):1353

[25] Lamichhane JR. Parsimonius use of pesticide-treated seeds: An integrated pest management framework. Trends in Plant Science. 2020;**25**:1070-1073. DOI: 10.1016/j.tplants.2020.08.002

[26] Jensen B, Knudsen IMB, Madsen M, Jensen DF. Biopriming of infected carrot seed with an antagonist, *Clonostachys rosea*, selected for control of seedborne *Alternaria* spp. Biological Control. 2004;**94**:551-560

[27] Okoth SA, Otadoh JA, Ochanda JO. Improved seedling emergence and growth of maize and beans by *Trichoderma harzianum*. Tropical and Subtropical Agroecosystems. 2011;**13**:65-71

[28] O'Callaghan M. Microbial inoculation of seed for improved crop performance: Issues and opportunities. Applied Microbiology and Biotechnology. 2016;**100**:5729-5746. DOI: 10.1007/s00253-016-7590-9

[29] Antoun H, Prevost D. Ecology of plant growth promoting rhizobacteria. In: Siddiqui ZA, editor. PGPR: Biocontrol and Biofertilization. Dordrecht: Springer; 2006. pp. 1-38

[30] Banerjee M, Yesmin L. Sulfur-Oxidizing Plant Growth Promoting Rhizobacteria for Enhanced Canola Performance. United State Patent number 20030172588 A1 and

WO2003057861, WIPO, Geneva, Switzerland. 2002

[31] Unno Y, Okubo K, Wasaki J, Shinano T, Osaki M. Plant growth promotion abilities and micro scale bacterial dynamics in the rhizosphere of lupin analysed by phytate utilization ability. Environmental Microbiology. 2005;7:396-404

[32] Bacon CW, White JF Jr. Physiological adaptations in the evolution of endophytism in the Clavicipitaceae. In: Bacon CW, White JF Jr, editors. Microbial Endophytes. New York, USA: Marcel Dekker Inc; 2000. pp. 237-261

[33] Kloepper JW, Mariano RLR. Rhizobacteria to induce plant disease resistance and enhance growth – Theory and practice. In: International Symposium on Biological Control for Crop Protection. Suwon, South Korea: Rural Development Administration; 2000. pp. 99-116

[34] El-Hendawy HH, Osman ME, Sorour NM. Biological control of bacterial spot of tomato caused by *Xanthomonas campestris pv. Vesicatoria* by *Rahnella aquatilis*. Microbiological Research. 2005;**160**:343-352

[35] Sundaramoorthy S, Balabaskar P. Biocontrol efficacy of *Trichoderma* spp. against wilt of tomato caused by *Fusarium oxysporum f. sp. lycopersici*. Journal of Applied Biology and Biotechnology. 2013;**1**:36-40

[36] Fatima K, Noureddine K, Henni JE, Mabrouk K. Antagonistic effect of *Trichoderma harzianum* against *Phytophthora infestans* in the north-west of Algeria. International Journal of Agricultural Research. 2015;**6**:44-53

[37] Ram LJ, Corrales DC, Soltani E. Biological seed treatments promote crop establishment and yield: A global meta-analysis. Agronomy for Sustainable Development. 2022;**42**:45. DOI: 10.1007/s13593-022-00761-z

[38] Cardarelli M, Woo SL, Rouphael Y, Colla G. Seed treatments with microorganisms can have a biostimulant effect by influencing germination and seedling growth of crops. Plants. 2022;**11**:259. DOI: 10.3390/plants11030259

[39] Jagadeesh V, Sujatha P, Triveni S, Keshavulu K, Jhansi Rani K, Raghavendra K. Effect of biological seed coating on Pigeonpea seedling vigour. International Journal of Current Microbiology and Applied Sciences. 2017;**6**(8):843-854

[40] Jagadeesh V, Sujatha P, Triveni S, Keshavulu K, Raghavendra K. Pigeonpea biological seed coating: Combined inoculation of biofertilizers and bioprotectants. International Journal of Current Microbiology and Applied Sciences. 2018;7(6):3692-3707

[41] Harman GE, Howell CR, Viterbo A, Chet I, Lorito M. *Trichoderma* species opportunistic, avirulent plant symbionts. Nature Reviews Microbiology. 2004;**2**:43-56

[42] Sawant I. *Trichoderma*-foliar pathogen interactions. The Open Mycology Journal. 2014;**8**:58-70. DOI: 10.2174/1874437001408010058

[43] Vinale F, Sivasithamparam K, Ghisalberti E, Woo S, Nigro M, Marra R, et al. Trichoderma secondary metabolites active on plants and fungal pathogens. The Open Mycology Journal. 2014;**8**:127-139. DOI: 10.2174/1874437001408010127

[44] Francesco VK, Sivasithamparam EL, Ghisalberti RM, Sheridan L, Woo ML. *Trichoderma* plant pathogen interactions.

Soil Biology and Biochemistry. 2008;**40**:1-10

[45] Limón MC, Chacón MR, Mejías R, Delgado-Jarana J, Rincón AM, Codón AC, et al. Increased antifungal and chitinase specific activities of *Trichoderma harzianum* CECT 2413 by addition of a cellulose binding domain. Applied Microbiology and Biotechnology. 2004;**64**(5):675-685. DOI: 10.1007/s00253-003-1538-6

[46] Sood M, Kapoor D, Kumar V, Sheteiwy MS, Ramakrishnan M, Landi M, et al. *Trichoderma*: The "secrets" of a multitalented biocontrol agent. Plants. 2020;**9**:762

[47] Murmanis L, Highley TL, Ricard J. Hyphal interaction of *Trichoderma harzianum* and *Trichodema polysporum* with wood decay fungi. Material und Organismen. 1988;**23**:271-279

[48] Ibrahim ME. In vitro antagonistic activity of *Trichoderma harzianum* against *Rhizoctonia solani* the causative agent of potato black scurf and stem canker. Egyptian Journal of Botany. 2017;**57**:173-185. DOI: 10.21608/ejbo.2017.903.1067

[49] Dubey SC, Suresh M, Birendra SS. Evaluation of *Trichoderma* species against *Fusarium oxysporum* f. sp. Ciceris for integrated management of chickpea wilts. Biological Control. 2007;**40**:118-127

[50] Khatabi B, Molitor A, Lindermayr C, Pfiffi S, Durner J, Wettstein D, et al. Ethylene supports colonization of plant roots by the mutualistic fungus *Piriformospora indica*. PLoS One. 2012;**7**:e35502

[51] Hussein SN. Biological control of root rot disease of cowpea *Vigna unguiculata* caused by the fungus *Rhizoctonia solani* using some bacterial and fungal species. The Arab Journal of Plant Protection. 2019;**37**(1):31-39. DOI: Doi.org/10.22268/AJPP-037.1. 031039

[52] Vidhyashekaran P, Kamala N, Ramanathan A, Rajappan K, Paranidharan V, Velazhahan R. Induction of systemic resistance by *Pseudomonas fluorescens* pf1 against *Xanthomonas oryzae* pv. Oryzae in rice leaves. Phytoparasitica. 2001;**29**:155-166

[53] Umesha S. Occurrence of bacterial canker in tomato fields of Karnataka and effect of biological seed treatment on disease incidence. Crop Protection. 2006;**25**:375-381

[54] Silva HSA, Romerio RDS, Macagnan D, Halfeld-Vieira BDA, Pereira MCB, Mounteer A. Rhizobacterial induction of systemic resistance in tomato plants: Non-specific protection and increase in enzyme activities. Biological Control. 2004;**29**:288-295

[55] Ji P, Campbell HL, Kloepper JW, Jones JB, Suslow TV, Wilson M. Integrated biological control of bacterial speck and spot of tomato under field conditions using foliar biological control agents and plant growth promoting rhizobacteria. Biological Control. 2006;**36**:358-367

[56] Guo JH, Qi HY, Guo YH, Ge HL, Gong LY, Zhang LX, et al. Biocontrol of tomato wilt by plant growth-promoting rhizobacteria. Biological Control. 2004;**29**:66-72

[57] Xiao K, Kinkel LL, Samac DA. Biological control of *Phytophthora* root rots on alfalfa and soybean with Streptomyces. Biological Control. 2002;**23**(3):285-295

[58] Gopalakrishnan S, Pande S, Sharma M, Humayun P, Keerthi Kiran BK, Sandeep D, et al. Evaluation

of actinomycete isolates obtained from herbal vermicompost for the biological control of Fusarium wilt of chickpea. Crop Protection. 2011;**30**:1070-1078

[59] Shahidi Bonjar GH, Barkhordar B, Pakgohar N, Aghighi S, Biglary S, Rashid Farrokhi P, et al. Biological control of *Phytophthora drechsleri* Tucker, the causal agent of pistachio gummosis, under greenhouse conditions by use of actinomycetes. Plant Pathology Journal. 2006;**5**:20-23

[60] Hamdali H, Hafidi M, Virolle MJ, Ouhdouch Y. Growth promotion and protection against damping-off of wheat by two rock phosphate solubilizing actinomycetes in a P-deficient soil under greenhouse conditions. Applied Soil Ecology. 2008;**40**:510-517

[61] Sadeghi A, Hessan AR, Askari H, Aghighi S, Shahidi Bonjar GH. Biological control potential of two Streptomyces isolates on *Rhizoctonia solani*, the causal agent of damping-off of sugar beet. Pakistan Journal of Biological Sciences. 2006;**9**:904-910

[62] Meschke H, Schrempf H. Streptomyces lividans inhibits the proliferation ofthe fungus *Verticillium dahliae* on seeds and roots of *Arabidopsis thaliana*. Microbial Biotechnology. 2010;**3**:428-443

[63] Haas D, Défago G. Biological control of soil-borne pathogens by fluorescent pseudomonads. Nature Reviews. Microbiology. 2005;**3**:307-319

[64] Paulitz TC, Belanger RR. Biological control in greenhouse systems. Annual Review of Phytopathology. 2001;**39**:103-133

[65] Walid N, Al-Jaramany L, Elbenay A, Al-Mhethawi R. Biological control of tomato damping-off and potato black scurf by seed treatment with *Trichoderma harzianum* Jordan. Journal of Biological Sciences. 2022;**15**(3):373-380. DOI: 10.54319/jjbs/150305

[66] Zhang Y, Li Y, Liang S, Zheng W, Chen X, Liu J, et al. Study on the preparation and effect of tomato seedling disease biocontrol compound seed-coating agent. Life. 2022;**12**(6):849

[67] Huang Y, Zhang X, Xu H, Zhang F, Zhang X, Yan Y, et al. Isolation of lipopeptide antibiotics from *Bacillus siamensis*: A potential biocontrol agent for *Fusarium graminearum*. Canadian Journal of Microbiology. 2022;**68**(6):403-411

[68] Akhil Reddy A, Sujatha P, Jhansi Rani K, Pushpavathi B, Rajeshwar Reddy T, Raghavendra K. Effect of seed coating types and storage period after coating on seed germination (%) and seedling vigour in Pigeonpea. Biological Forum – An International Journal. 2022;**14**(1):1328-1335

[69] Koohakan P, Prasom P, Sikhao P. Application of seed coating with endophytic bacteria for Fusarium wilt disease reduction and growth promotion in tomato. International Journal of Agricultural Technology. 2020;**16**:55-62

[70] Rajput RS, Singh J, Singh P, Vaishnav A, Singh AB. Influence of seed biopriming and vermiwash treatment on tomato plant's immunity and nutritional quality upon *Sclerotium rolfsii* challenge inoculation. Journal of Plant Growth Regulation. 2020;**40**:1-17

[71] Bahramisharif A, Rose LE. Efficacy of biological agents and compost on growth and resistance of tomatoes to late blight. Planta. 2019;**249**:799-813. DOI: 10.1007/s00425-018-3035-2

[72] Biam M, Majumder D. Biocontrol efficacy of *Trichoderma* isolates against

tomato damping-off caused by *Pythium* spp. and *Rhizoctonia solani* (Kuhn.). International Journal of Chemical Studies. 2019;7(3):81-89

[73] Tekale AG, Guldekar DD, Kendre VP, Deshmukh AP, Potdukhe SR. Efficacy of fungicides and bioagents against damping-off in Chilli caused by *Pythium aphanidermatum* International Journal of Current Microbiology and Applied Sciences. 2019;**8**(6):637-648. DOI: 10.20546/ijcmas.2019.806.074

[74] Uddin MN, Rahman UU, khan W., et al. Effect of *Trichoderma harzianum* on tomato plant growth and its antagonistic activity against *Phythium ultimum* and *Phytopthora capsici*. The Egyptian Journal of Biological Pest Control. 2018;**28**:32. DOI: 10.1186/s41938-018-0032-5

[75] Sain SK, Pandey AK. Biological spectrum of *Trichoderma harzianum* Rifai isolates to control fungal diseases of tomato (Solanum lycopersicon L.). Archives of Phytopathology and Plant Protection. 2016;**49**:507-521. DOI: 10.1080/03235408.2016.1242393

[76] Kipngeno P, Losenge T, Maina N, Kahangi E, Juma P. Efficacy of *Bacillus subtilis* and *Trichoderma asperellum* against *Pythium aphanidermatum* in tomatoes. Biological Control. 2015;**90**(2015):92-95

[77] Hassan WA, Taher IE, Saido KA, Ali AS. Antagonistic succession of *Trichoderma* against Rhizoctonial damping- off on tomato in composted media. Bulletin of the Iraq Natural History Museum. 2015;**13**(4):41-49

[78] Chohan S, Perveen R, Mehmood MA, Naz S, Akram N. Morpho-physiological studies, management and screening of tomato germplasm against *Alternaria solani*, the causal agent of tomato early blight. International Journal of Agriculture and Biology. 2015;**17**:111-118

[79] Goudjal Y, Toumatiaa O, Yekkour A, Sabaoua N, Mathieuc F, Zitouni A. Biocontrol of *Rhizoctonia solani* damping-off and promotion of tomato plant growth by endophytic actinomycetes isolated from native plants of Algerian Sahara. Microbiological Research. 2014;**169**:59-65

[80] Ajilogba CF, Babalola OO. Integrated management strategies for tomato *Fusarium* wilt. Biocontrol Science. 2013;**18**(3):117-127

[81] Loliam B, Morinaga T, Chaiyanan S. Biocontrol of *Phytophthora infestans*, Fungal Pathogen of Seedling damping-off Disease in Economic Plant Nursery. Hindawi Publishing Corporation Psyche. New York; 2012. DOI: 10.1155/2012/324317

[82] Hamed HA, Moustafa YA, Abdel-Aziz SM. In vivo efficacy of lactic acid bacteria in biological control against *Fusarium oxysporum* for protection of tomato plant. Life Science Journal. 2011;**8**(4):462-468

[83] Veerubommu S, Nandina K. Biological management of vascular wilt of tomato caused by *Fusarium oxysporum f.sp. lycospersici* by plant growth-promoting rhizobacterial mixture. Biological Control. 2011;**57**(2011):85-93

[84] Someshwar B, Sitansu P. Biopriming of seeds for improving germination behavior of chilli, tomato and brinjal. The Journal of Mycology and Plant Pathology. 2010;**40**(3):375-379

[85] Mansouri F, Bjorkman T, Harman GE. Seed treatment with with *Trichoderma harzianum* alleviates biotic, abiotic and physiological stress

in germinating seed and seedling. Phytopathology. 2010;**100**:1213-1221

[86] Chandel S, Allan EJ, Woodward S. Biological control of *Fusarium oxysporum f. sp. lycopersici* on tomato by *Brevibacillus brevis*. Journal of Phytopathology. 2010;**158**(7-8):470-478

[87] Rashmi S, Khalid A, Singh US, Sharma AK. Evaluation of arbuscular mycorrhizal fungus, fluorescent pseudomonas and *Trichoderma harzianum* formulation against *Fusarium oxysporum f. sp. Lycopersici* for the management of tomato wilt. Biological Control. 2010;**53**:24-31

[88] Mastouri F, Björkman T, Harman GE. Seed treatment with *Trichoderma harzianum* alleviates biotic, abiotic, and physiological stresses in germinating seeds and seedlings. Phytopathology. 2010;**100**:1213-1221. DOI: 10.1094/PHYTO-03-10-0091

[89] Srivastava R, Khalid A, Singh US, Sharma AK. Evaluation of arbuscular mycorrhizal fungus, fluorescent pseudomonas and *Trichoderma harzianum* formulation against *Fusarium oxysporum f. sp. Lycopersici* for the management of tomato wilt. Biological Control. 2010;**53**:24-31

[90] Bennett AJ, Mead A, Whipps JM. Performance of carrot and onion seed primed with beneficial microorganisms in glasshouse and field trials. Biological Control. 2009;**51**:417-426

[91] Pill WG, Collins CM, Goldberger B, Gregory N. Responses of non-primed or primed seeds of 'Marketmore 76' cucumber (*Cucumis sativus* L.) slurry coated with *Trichoderma* species to planting in growth media infested with *Pythium aphanidermatum*. Scientia Horticulturae. 2009;**121**:54-62

[92] Kavitha R, Umesha S. Prevalence of bacterial spot in tomato fields of Karnataka and effect of biological seed treatment on disease incidence. Crop Protection. 2007;**26**:991-997

[93] Zaghloul RA, Hanafy Ehsan A, Neweigy NA, Khalifa NA. Application of biofertilization and biological control for tomato production. In: 12th Conference of Microbiology; March 18-22. Cairo, Egypt: Benha University; 2007. pp. 198-212

[94] Morsy EM, Abdel-Kawi KA, Khalil MNA. Efficiency of *Trichoderma viride* and *Bacillus subtilis* as biocontrol agents against Fusarium solani on tomato plants. The Egyptian Journal of Phytopathology. 2009;**37**:47-57

[95] Montealegre JR, Reyes R, Pérez LM, Herrera R, Silva P, Besoain X. Selection of bioantagonistic bacteria to be used in biological control of *Rhizoctonia solani* in tomato. Electronic Journal of Biotechnology. 2003;**6**(2):115-127

[96] Warren JE, Bennett MA. Bio-osmopriming tomato (*Lycopersicon esculentum* mill.) seeds for improved seedling establishment. Seed biology: Advances and applications. In: Proceedings of the Sixth International Workshop on Seeds, Merida, Mexico. Wallingford, UK, New York: CABI Pub.; 2000

[97] Weller DM. Bio control of soilborne plant pathogens in the rhizosphere with bacteria. Annual Review of Phytopathology. 1988;**26**:379-407

[98] Sivan A, Ucko O, Chet I. Biological control of *Fusarium* crown rot of tomato by *Trichoderma harzianum* under field conditions. Plant Disease. 1987;**71**(7):587-592. DOI: 10.1094/PD-71-0587

Section 2

Agronomic Practices

Chapter 4

Greenhouse Tomato Technologies and Their Influence in Mediterranean Region

Raquel Saraiva, Igor Dias, José Grego and Margarida Oliveira

Abstract

Tomato (*Solanum lycopersicum* L.) is the most consumed vegetable and one of the most studied crops in the world. Over the years, several technologies have been studied and applied to crop production towards higher productivity, quality, and production efficiency. This chapter reviews greenhouse tomato production, cropping systems, and environmental conditioning, focusing on technological developments and the latest reclaimed water trends that have started to take off in the context of increasing water scarcity due to climate change. Following worldwide research trends and policies, the influence of the different technologies in fresh tomato production and the use of reclaimed water or reuse of treated nutrient solution is explored as it is expected to be a great advance in the Mediterranean region in the next years, and it is of the utmost importance, as the region increasingly suffers from climate change effects.

Keywords: horticultural practices, hydroponics, nutrient recovery, reclaimed water, *Solanum lycopersicum* L

1. Introduction

The United Nations (UN) Sustainable Development Goals aim to ensure sustainable food production systems and the implementation of resilient agricultural practices that increase productivity and production by 2030. At the same time, it is required to achieve sustainable consumption and production patterns [1].

Tomato (*Solanum lycopersicum* L.) is one of the most consumed vegetables in the world due to its nutritional characteristics and benefits to health [2, 3]. In 2021, world tomato production was around 189,133 million tonnes, with China as the country with the highest world production, followed by India and the Mediterranean basin, respectively [4]. In the report "The 2017 Agricultural Outlook Conference," EU forecasts indicated that fresh tomato production would remain stable despite the expected increase in productivity caused by wider production seasons [5]. Rising temperatures and extreme weather conditions, changing rainfall and snow cover patterns, as well as the increase in the frequency of floods and droughts due to climate change, are a challenge for the agricultural sector [6]. These factors, allied with the growth in world's population and the need for food availability, are the driving force behind the sector's progress. Exportation, consumer demand for quality and availability, competitively with other markets and climatic challenges influence tomato production around the world and technological solutions, help producers to face these demands [6]. Greenhouses for tomato production can be a means to increase productivity with efficient use of water, nutrients, and energy. In order to achieve optimal production conditions for both daytime and night-time temperatures, there is a significantly high energy consumption which can be overcome by adopting new climate control approaches to substitute the use of conventional energy, with cogeneration or renewable energy sources [7]. Nowadays, there are generically two types of greenhouses, in different geographic regions, with different degrees of sophistication and applied technologies, depending on climatic and socio-economics conditions: Mediterranean basin countries use typically low-tech, plastic-covered greenhouses; Northern Europe and the United States use glass greenhouses with more technologic solutions [8]. Each degree in complexity induces a new degree of microclimatic modification and adds new layers of benefits: high-tech greenhouses are expensive and involve high initial investment but lead to higher yields [7]. On the other hand, low-tech, plastic greenhouses, with mostly natural ventilation, are a low-cost solution that enables higher productions and wider production seasons than an open-field, with low initial investment [7]. However, the depletion of solar UV in greenhouse crops might compromise the sensorial perception of the fruits compared to open-field ones [3, 9]. Some events can boost the evolution of greenhouses, as is the case of west Portugal, increasing innovation as the greenhouses have been substituted and the brands expand their activity [10]. In this case, the boost of technological advances arrived with a natural disaster in 2009 that shredded several structures and forced the producers to rebuild, with State support, in an opportunity to rethink, improve and adopt more sustainable structures to achieve the best yields with the lowest environmental impacts [11].

In recent years, new technologies for greenhouse production have been developed to improve efficiency in water, energy, and chemical use, as well as fruit quality and shelf life [12]. Nevertheless, experience shows that the transfer of technologies and practices from North Europe, as they are more sophisticated and high-tech, to realities with different climatic conditions and socio-economic environments does not bring good results. Technology transfer should be adapted, tested, and validated to local requirements [8]. Therefore, crop management, technics and technologies have been tested. Some of them had success and were implemented, but others were left behind [8].

2. Varieties, grafting and rootstocks

New cultivars are commercialized every year, mostly hybrids, and their choosing, along with fruit number per cluster, affects fruit weight and disease resistance [13].

The European's Court of Justice decision of July 2018 is being contested by hundreds of European researchers asking EU institutions to revert the decision of including genome-edited organisms, with no foreign DNA, in the restricted legislation for Genetic Modified Organisms (GMO), but instead fall under the conventionally bred varieties regime under penalty that, if not, Europe's science and agriculture fall behind the rest of the world in such crucial tool for production in the face of climate change [14, 15].

Grafting is widely used as a rapid tool to increase environmental stress tolerance on crops, and currently, most crops are grown as scion-rootstock arrangements

[16]. Although this technique requires more labour for breeders, it promotes resistance to biotic and abiotic stresses, promotes more vigorous plats or higher yield, and improves fruit quality [7, 16, 17]. Since rootstocks can affect nutrient uptake, selecting rootstocks with higher nitrogen acquisition ability lead to a decrease in N fertilizer application without affecting tomato yields, while minimizing environmental pollution [18]. The use of highly resistant transgenic plants as rootstocks and grafted onto non-transgenic tomato scion result in non-transgenic plants that may evade the concern of GMO biosafety regulation, while successfully transferring resistance to the scions [17]. On the contrary, some reports state that in the absence of any induced stress, rootstocks have no effects on crop growth and water relations [19] and, under optimal conditions, some rootstocks may even reduce the growth of the grafted plant [16].

Despite the several studies available, tomato growers are not persuaded to prune clusters to a determined number or use reports as a base because of the conflicting results they present [20]. Instead, they go by the requirements of the market, the length of the cycle and their own experience [6], preferring to test on a small scale on their own before applying any of the suggested techniques. In a region most affected by climate change, all the biotechnology must be used to fight its effects.

3. Production systems

Tomato can be planted on substrates or on soil; in this last case, most producers use mulching plastic films [21]. Greenhouse provide ideal environmental conditions for the development of diseases; in addition, soil production brings some challenges as salinization and soil diseases resulting from the lack of crop rotation [22]. Chemical soil disinfection is the primary method used to manage soil-borne plant pathogens, reduce nematode populations, control soil sickness and allow profitable production [23, 24]. Soil solarization was introduced in the 1970s and is used worldwide as an environmentally friendly alternative to chemical disinfestation, with proven results in inducing physical, chemical and microbiological modifications in the soil, stimulating root growth and crop yield [25], having the advantage of non-toxicity, that is one of the greatest concerns related to other treatments.

Another string of action against the use of chemical pesticides is the implementation of different pest management strategies such as the use of intercrop plants with natural pesticides proprieties [23]. Improving soil quality is another challenge in this cultivation system, since it is critical for fertility and crop health [26]. The addition of compost for increasing soil organic matter and providing nutrients for the crop or the addition of biochar as soil amendment can increase soil sustainability and crop productivity by the increase on carbon sequester and nitrogen retention [26]. Combining soil solarization with organic fertilization techniques was proven to maximize their effects, achieving higher soil temperatures and release of volatile compounds, resulting in higher mineral availability and improved yields [27, 28].

Over the years, greenhouse soil production has been gradually substituted by hydroponic soilless production, with Nutrient Film Technique (NFT) systems or, more usual, on sacks of substrate (rockwool, coconut fiber, perlite or vermiculite) [29, 30] that provide a much uniform medium and allow more control of production conditions. Several substrates were studied over the years, but only the most stable ones remained. In Europe, greenhouse tomatoes are mostly produced in substrate systems, mainly rockwool [30], with individual drippers to plant-by-plant irrigation [29]. All plant nutrients are supplied constantly in the irrigation water, and the fertilizer mixing, pH and electrical conductivity (EC) are continually monitored and can be electronically controlled. By being soilless, hydroponic systems protect crops from phytosanitary problems by starting out clean [29]. Rockwool is made from rock fiber that can be steam-sterilized and reused and provides roots with the ideal environment to grow [31]. The slabs can be used for a second, and sometimes a third, crop. Prior to reuse, the slabs are sterilized, and new plastic sleeves are placed around the slabs [13]. Growing environmental concerns about the amount of rockwool waste generated led to some attempts to find suitable substrate substitutes: organic or inorganic substrate, with one component or mixtures (sheep wool, cultivated Sphagnum biomass, hemp sawdust, coco coir, volcanic sand, perlite) [20, 32].

NFT system, on the other hand, consists of growing plants without the use of a substrate, but instead, conserving a shallow nutrient solution (NS) around the roots and can improve yield up to three times, depending on NS concentration [33]. In [34] the cultivation of tomatoes in soil was compared with NFT production (closed cycle) and rockwool (open cycle). The results show that hydroponic systems enhanced earliness in at least 8 days and yield in at least 10 t/ha compared to soil. Higher yield, less water and fertilizer application were achieved in NFT, thus reducing the environmental impact.

4. Microorganisms/inoculation

The introduction of specific microorganisms into the crop system result in several benefits such as improving nutrient use efficiency, plant growth, fruit weight and disease control [35, 36]. Studies in plant inoculation with bacteria and fungi demonstrate the positive effects of *Pseudomonas putida*, *Trichoderma atroviride* or *Methylobacterium* in promoting the reproductive growth of tomatoes under hydroponic growing systems, improving yields by >15% in organic medium [35, 37, 38]. Bacteria of genus *Methylobacterium* have been found to significantly reduce disease symptoms and lower the disease index value of *Ralstonia solanacearum* in inoculated plants when compared to non-inoculated ones, suggesting its potential use as a biocontrol agent in tomatoes [38]. Substantial post-harvest disease control can also be accomplished by inoculation (*Bacillus subtilis* strain QST 713) when combined with storage at 13°C [36].

5. Pollinators

Bumblebees (*Bombus* spp.) are commonly used for pollination in greenhouse, being the principal pollinator in tomatoes [39]. Its use is very effective and economically profitable, inducing greater fruit sets, more regular form, more top-grade fruits, and lowest defects [40, 41]. Besides pollination, [42] conducted a study to assess the contribution of bumblebees for the dissemination of two fungi, in greenhouse tomato and sweet pepper, for control of pests and plant disease. The authors demonstrate that pollinators can influence pests and disease control since the simultaneous spread of *Beauveria bassiana* and *Clonostachys rosea* inoculums by bumblebees contributed to controlling greenhouse whitefly, reducing plant pests, and gray mold (*Botrytis cinerea*), having no effect on bee mortality.

Using commercial bumblebees raises some ecological concerns and real problems that have not been addressed: as colonies are transported all over the world, subspecies are introduced into non-endemic regions, there is competition between species, genetic pollution of native species, parasites and pathogens transmission and changes in the native flora [41]. Nevertheless, bumblebees used for pollination are a better choice compared to plant growth regulators that could also present environmental hazards and furthermore, growers have more careful in using chemicals in the greenhouse when using bumblebees [43]. To minimize the problems, hive boxes should be disposed of safely. The development of commercial native bumblebees must be investigated, and legislation should address and regulate its use [43].

6. Greenhouse environment

Greenhouse environment can be altered by several factors as the choice of cover materials, lightening, heating or cooling, aeration, CO_2 application, the use of evaporative systems and others.

6.1 Greenhouse materials

Covering materials increase fruit yield and quality by influencing the environment inside, such as thermal conditions [44, 45]. Solar radiation is the main factor affecting plant photosynthesis and, consequently, plant growth [46]. Ultraviolet radiation (UV) depletion affects the sensorial and nutritional quality of tomatoes, and off-season fruits are less tasteful than in-season fruit, and these are usually less tasteful than open-field ones [9]. As a result, greenhouse cladding materials characteristics are very important, as they affect transmitted radiation, reflectivity, and absorptivity [45–48], leading to different interior environmental conditions (with consequences on air temperature, relative humidity, and vapor pressure deficit (VPD)) [49], affecting physiological processes and yield, remaining a primary concern in greenhouse production [48, 49]. An appropriate covering material is assumed to combine low transmittance in the long wave band with the maximum transmittance in the photosynthetically active radiation (PAR) spectrum [50].

Greenhouses provide a better environment for plants to grow in cold climates by conserving heat, but in hotter climates, like the Mediterranean, cooling systems need to be used to lower temperature and increase humidity to provide an adequate environment [51]. The inefficiency of greenhouse covers, such as low visible light transmittance and low near-infrared reflectance [52], led to the need for heating or cooling greenhouses, which are very energy-consuming processes. To reduce energy consumption, in the last decades, the use of thermal-resistant materials, curtains, alternative sources of energy and techniques to reduce energy loss gained huge importance [45].

The thickness of the cladding material, dirtiness, the use of additives for the radiation spectrum and the condensation on the interior affect radiation transmittance through the covering material [44, 47, 53]. Plastic materials are the most popular cover mainly by the use of low-density polyethylene (LDPE), the copolymer of ethylene and vinyl acetate (EVA) and polyvinylchloride (PVC) [45, 48, 54]. Polyethylene films are mostly LDPE due to their relatively good mechanical and optical properties (flexible, transparent to thermal radiation and with 70–95% solar transmittance) combined with a competitive market price [48]. Polycarbonate covers are more affordable than glass and have good thermal efficiency but avoid the transmittance of UV light, allowing only 400 nm or higher wavelengths to reach the fruits [55]. UV-A/B radiation, control several plant functions, and a tomato grown under reduced UV radiation has its quality compromised [3]. By choosing covering materials with higher UV-transmittance in tomato production, the anti-oxidative capacity of fruits can be improved without influencing fruit weight [56].

Tantau et al. [47] tested 20 different covering materials and measured PAR and solar radiation, and compared them to the outside conditions. The results corroborate the early ones and stated that condensation on the internal surfaces of the cladding materials reduces PAR transmittance. During the 5 years, the ageing effect on the material's light transmission was not detected. Covers, structures, and fixed objects hanging over the plants can block radiation in 25 to 50% [50].

Shade nets can protect greenhouse crops from high light intensities, and its use is a simple and very effective, strategy for cooling greenhouses during the summer [50, 57]. In Greece, [58] tested the effects on microclimate, crop production and quality of tomato grown under four different shade nets, finding that shading does not reduce dry matter content and, on the other hand, increase leaf area index, the number of fruits per plant, and the total fresh tomato yield. Tomato cracking was reduced by 50% in shading, and marketable tomato fruit yield increased by around 50% compared to non-shaded treatments.

In [44], tomato color development and lycopene accumulation were assessed in fruits grown in a) double-layer polyethylene with infrared transmission and anti-condensate (IR/AC) additives and b) flat glass 4 mm thick coated with a 15% $CaCO_3$ solution greenhouses. Their results suggest that the lycopene biosynthesis process is affected by the season of production, temperature, and lighting conditions and in the 32nd week, lycopene content was higher in the fruits collected in a double layer of polyethylene greenhouses. The application of 15% $CaCO_3$ solution helped to control the temperature but had negative effects on lycopene biosynthesis. Flat glass with a $CaCO_3$ coating showed higher light transmission and, between 380 and 600 nm wavelength, the glass coating allowed higher light transmission than the double layer of polyethylene, but at 650–760 nm, plastic cover presented higher light transmission. Pigmented coatings are found to reflect incident solar radiation in the infrared wavelengths (NIR), while maintaining visible light (VIS) to be transmitted and reducing heat. It is possible to have higher transmittance in VIS than NIR using synthetic diamond particles with the low optimum volume fraction, 0.89%, in powder form and with nearly 1.19 μm as optimum diameter, and so it could be used as a pigment material for reducing cooling energy requirements in hot climates [51]. Colored paints are used for cooling as well. [59] studied coated flakes of metal to achieve enhanced solar reflectance and lower inside temperatures. They found that flakes of metal coated with a single layer of iron oxide or a double layer of iron oxide on silica (<200 nm thickness) have lower solar absorbance than conventional paints.

Public health concerns regarding chemical pesticide toxicity restrict their use, but they are still very important in agriculture, so another innovative approach to cover materials was made [60], where nanocomposite film was used as a pesticide delivery system. The authors loaded deltamethrin into halloysite nanotubes, which resulted in improvements in the flexible modulus of linear LDPE films and obtained continuous release for 60 days. The results revealed efficacy in repelling mature aphids and eradication of young aphids and thrips when placed at 1 m of plant's boundary.

6.2 Lightning

As light is a major factor in plant growth and its insufficiency is a key abiotic stress, limiting plant development and yield [61], several studies try to get insight into the influence of additional light sources, the extension of photoperiods, spectrum ranges and/or complementary light effect on plant development [50, 61–65]. The extension of photoperiods has been studied for more than three decades to improve leaf fresh weight and tomato yields and to support year-round production in greenhouse horticulture where natural light is limited [65, 66]. In the early studies, the common source of supplementary light were high-intensity discharge light sources such as metal halide lamps and high-pressure sodium (HPS) lamps, but nowadays, light-emitting diodes (LED) are the most common technology used [65]. High-intensity discharge light systems are very inefficient and produce high amounts of radiant heat [65]; LEDs, on the other hand, are an energy-efficient alternative since it converts electricity to light much more efficiently [65, 67] and can be designed to fit the spectral demand of the plants [61, 65].

All wavelengths proved to affect crops positively, but different light spectrum wavelengths induce specific effects on plants [68]. Blue light (450–495 nm) is reported as a chlorophyll formation regulator and growth promotor [69, 70]. There is a generalized perception that green light does not contribute or is less effective in growth than other ranges of the visible spectrum [71], but contradictory results can be found in the literature on its benefits and disadvantages. Several studies in lettuce claim that enriching a white or red/blue spectrum with green light (while keeping light intensity identical) can increase plant growth and yield [72]. At the same time, others reported that green light (510–585 nm) penetrates the canopy more than red or blue ranges [71], improving the vertical light profile and the irradiance in otherwise shaded leaves. So, it can be hypothesized that green light may have some importance in plant growth [73].

Red light (600–740 nm) is important for the development of the photosynthetic system and is often associated with the morphological adjustment of plants to the environment [69]. It is reported that red lightning environments impose physiologic responses and affect floral initiation, chlorophyll, and carotenoid contents [68]. A prolonged period of monochromatic red-light treatment has shown positive effects on the yield and quality of greenhouse tomatoes [68, 70].

According to [69], plants respond to UV-A treatments through cryptochromes and phototropins and under UV-A/B irradiation treatments, plants presented enhanced flowering, fruit ripening synchronization and fruit's nutritional properties as fruit °Brix and other physicochemical characteristics [3]. Combinations of a specific wavelength from LEDs appears to enhance nutritional and health characteristics of horticultural crops [56, 66]. In fact, the effects of supplemental lighting with three different lighting source approaches on the dynamics of fruit growth and composition in soluble sugars, starch and acids in tomatoes were studied by [65] and reinforced the results of [63] with the discovery that regarding the application of a low supplementary dose of far-red LED light to red/blue light, fruit quality is improved: glucose, fructose and sucrose were significantly increased in pericarp, having a great impact in taste characteristics of tomato [74]. When applying intracanopy red + far-red LED light, fruit yield can be 50% higher compared to the use of HPS lamps while improving fruit quality and of-season tomato flavor, reducing energy use by 70% [75].

In [64], light-induced change in tomato metabolic processes morphology and production response to exposure to three durations of far-red (FR) supplemental lighting is investigated, combined with red and blue lightning. The results show that supplemental FR light resulted in longer plants, with a higher leaf length/width ratio and larger leaf area compared with the control, which led to a more homogeneous intracanopy light distribution. Contradicting the results of [65], in [64] reported a certain decrease in soluble sugar content of the ripe tomato fruit even though a 7–12% ripe fruit increment was achieved. With no significant differences, 0.5 h treatment was the most favorable, since it stimulates growth and production, as the others, but is less energy consuming.

The application of supplemental light within the lower canopy or inter canopy could be a better way to suppress light insufficiency than traditional supplemental light [61]. LEDs permit placement closer to crop canopies due to the cool surfaces, which greatly reduces electrical energy requirements. Light distribution can be enhanced while decreasing the waste of light that other technologies present and spectral blends can be tailored for each purpose [66]. LED interlighting in tomatoes resulted in increases irradiation levels [61] and yield by 24–36% [76]. Li et al. [77] related the application of interlighting with higher water use efficiency in tomatoes. Supplemental lighting provided from underneath the canopy (UC) or within the inner canopy (IC) in single-truss tomato plants significantly promoted leaf photosynthetic activities, plant growth, and fruit production in plants exposed to low solar irradiation levels [61]. The use of UC had a greater contribution to fruit yield; however, IC induced an increase of soluble solid contents in fruit due to the higher exposure of the fruit to direct supplemental light. Moreover, the cost–benefit analysis shows that the application of UC may be more economically attractive than the use of IC. The application of narrow-band red and blue light as intracanopy lightning did not have any negative effects on fruit quality, and, with solely white LED interlighting, positive effects were demonstrated for tomatoes [78]. LED lighting may also be a suitable way to control pest insects and prevent physiological diseases stimulated by low-intensity or narrow-spectrum [66].

The influence of low irradiance light pulses (LP) of 15 min each was tested during the night, with a frequency of 2 h and 4 h, applied to plants in a greenhouse with controlled temperature after fruiting until ripening in red [79]. Results demonstrated that low irradiance LP treatment reduces the concentration of free sugars, amino acids, and other metabolites without impact on other fruit quality parameters such as firmness. Plants exposed to 15 min LP every 2 h during the night presented an increased yield of 18%.

6.3 Energy requirements

Intensive greenhouse energy requirements lead to important energy costs that could easily constitute 15 to 30% of a greenhouse operation's annual costs [80] and, in some cases, could even reach 50% of the total annual cost [81]. Nowadays, with growing climate change concerns, and actual challenges, limitations in fossil energy supply and heating costs rising for conventional coal, oil and natural gas, the use of energy conservation techniques and innovative applications of renewable energy have gained increased importance to heat or cool the greenhouses, while improving energy utilization, contributing directly to the reduction of greenhouse gas (GHG) emissions and aiding to reduce production costs [82, 83]. Thus, greenhouse systems based on solar energy to increase air temperature or for energy production seem very appealing and have been studied along the Mediterranean basin. Among them, it could be found: rock-bed storage, water storage, movable insulation, ground air collectors, phase change material storage, north wall storage and the installation of photovoltaic panels [83]. These systems are capable, to more or less extent, of increasing the greenhouse air temperature during the night compared to the conventional greenhouse and increasing yields [84, 85] and the application of a passive solar system like polyethylene sleeves along rows can lead to 8% energy saving in heated greenhouses [86] while the use of north wall insulation systems can achieve 31.7% reduction in heating requirements [87].

Using semi-transparent build-in integrated photovoltaics mounted on 20% of the greenhouse roof area, [88] found no significant differences in tomato development compared to control even though a 35 to 40% reduction of solar radiation was observed, resulting in 1 to 3°C decrease in air temperature on clear days, not affecting relative humidity.

Other renewable energy sources as small/medium-sized wind turbines and biomass, could also be explored to reduce energy costs in heated greenhouses [89]. Tomato crop is negatively affected by low temperatures; however, a balance between energy use and crop production can be achieved. Reducing heating set points during the night could represent a huge energy-saving measure without affecting total yield, but loss of early production profit due to the delay of harvest has to be weighted by producers [90].

6.4 Environmental control

High humidity in a greenhouse is favorable to fungal disease development, so its control is of extreme importance. Ventilation has a very important contribution in reducing high relative humidity, minimizing diseases and, consequently, minimizing the use of chemicals in greenhouses [91]. Natural ventilation depends on the wind and temperature differences between outside and inside, rely on doors, openings on the roof and/or in the side walls [91], and uses no energy but is usually insufficient. Forced ventilation for cooling and air humidity regulation purposes, and mostly the fan-pad system, is widely used worldwide to control high temperatures in summer, provide uniform airflow, and control CO_2 concentrations levels [92]; however, it also comes with an energy cost. To reduce the costs, natural ventilation combined with mechanical cooling could be used [93], but also, the maintenance of equipment is of great importance [94].

Aeration practices are proven to increase tomato yield and fruit quality but also to intensify soil CO_2 and N_2O emissions without, however, significantly increasing the overall greenhouse [95–98]. Some studies indicate the existence of complex interactions among growing seasons, irrigation method, and N application, which indicate that in increasing global warming scenario, using aerated irrigation in combination with reducing nitrogen fertilizer rate could be an effective way of maintaining crop yield while reducing soil net GHG emissions [98].

Low carbon dioxide is one of the primary factors affecting the quality of greenhouse tomatoes [99]. In general, the net photosynthetic rate increases with increasing CO_2 concentration, 800–1000 $\mu L\ L^{-1}$ considered the optimal CO_2 concentration for plant development [100]. Thus, CO_2 enrichment in controlled environmental conditions is known to effectively promote photosynthesis, enhance growth and production, increase water use efficiency (WUE) and is widely used around the world [101]. The contents of lycopene, β-carotene, ascorbic acid, sugars, titrable acidity and sugar/acid ratio were increased in fruits grown in CO_2 enrichment environment, and organoleptic characteristics were beneficiated [99].

For energy and environmental sustainability, alternatives to traditional CO_2 enrichment devices are being studied, and [101] suggested the symbiosis between industrial installations (production of CO_2) and greenhouses to reduce the overall amounts of CO_2 released in the atmosphere, recommending, however, that a case-by-case analysis be conducted to assess the environmental benefit and economic feasibility.

To achieve desired growing conditions in summer, the control of air temperature and VPD by fogging technology is intensifying. This technology has been reported as useful in reducing the need for natural ventilation, allowing for higher CO_2 concentrations to be maintained inside greenhouses and can influence the number and mass of the fruits [102, 103]. Evaporative fogging systems are considered better than pad–fan systems for achieving cooling and increasing absolute humidity in a more uniform spatial distribution, providing higher water evaporation while keeping the plants dry [104], although its efficiency depends on system design parameters and conditions. However, operation costs and water availability must be assessed prior to the installation of such systems [103].

Each environmental control technology and crop management strategies have huge impacts in greenhouse production. The interactions between those two vectors (**Table 1**) enhance the beneficial effects, resulting in more thoughtful greenhouses, more efficient, productive, and capable of contributing to sustainable production in the face of climatic change and world challenges for feeding the increasing population.

Greenhouse environmental control systems are becoming more sophisticated, using more and more sensors to improve control, and as sensor prices are reduced due to development and higher offer, they are being used more frequently [105]. These systems can collect outside and inside data about temperatures, wind direction, sunlight, carbon dioxide levels, and humidity, among other parameters, to better control the greenhouse environment for optimizing cultivating conditions: opening and closing windows, adjusting NS, start/stop forced ventilation or heating to obtain the desired temperature, humidity, light, pH, EC, water, and nutrition levels [105, 106].

Automation programmes, timers, temperature, humidity, pH, CE and other sensor are widely used to control fertigation [107, 108]. Other them that, recent developments in automation and robotics aim to increase production efficiency in greenhouses by studding solutions for labour-demanding tasks [7]. Most of the studies present solutions for harvesting robots with successful results [109–111]. The fruit's physical and mechanical properties were studied by [112] to allow automated harvest and other studies are reported in tomato identification by the artificial vision-robotic system [113]. A pesticide robot spray application was tested in tomatoes that provided encouraging results with the same or improved performance in comparison to traditional pesticide application methods [114]. Besides mechanical harvesting, a Greenhouse Robotic Worker - GRoW - can perform various tasks, like, trimming, monitoring the crop and pollinating, estimating a reduction of labour costs up to 50% [110].

7. Irrigation techniques and reclaimed water trends

Drip fertigation is one of the best techniques for applying water and fertilizer to fruit and vegetables and has been widely used in greenhouses [115, 116] associated with automation programmes based on timers, sensors and models or manual

		Irradiation	RH	VPD	Temp.	CO_2	Air Flow	Heating Req.	Emissions	Yield	Quality	Shelf Life	Diseases
Covers	Plastic	+						—					
	Glass	+						+		+	+		
	Additives	+			+						+		
	Nets	—			+					+	+		
	$CaCo_3$	—	+		+						—		
Complemental light	Extent Photoperiod									+	+		
	Far-Red										+		
	Red									+	+		
	Blue									+			
	Green									^			
	UV-A/UV-B									+	+	+	
	Interlighting									+	+		
	Under Canopy									+			
	Pulses									+			
Energy	Heating				+					+			—
	Cooling				+					+			+
	Ventilation		+		+	+	+						+
Other environment	Evaporative Systems			+	+	+				+			
	CO_2									+	+		
	Aeration								—	+	+		

		Irradiation	RH	VPD	Temp.	CO_2	Air Flow	Heating Req.	Emissions	Yield	Quality	Shelf Life	Diseases
Variety/ Grafting/ / Rootstock	Adequate Cultivar									+	+	+	+
	Grafting & rootstock									^	^		+
Medium	Soil									—			—
	Substrate									+			+
	NFT									+			+
Others	Microorganism inoculation									+		+	+
	Pollinators									+	+		+

Table 1.
Summary of influence for greenhouses. (+) positive effect; (−) negative effect; (^) conflicting results.

activation [107, 108]. Negative pressure irrigation is also a commonly used irrigation technic in greenhouses [117]. It supplies water at a negative pressure, controlling soil water content, allowing automatic irrigation according to crop requirements [118], can increase yield, fruit height, stem diameter, and significantly improve water use efficiency (WUE) when compared to drip irrigation [116]. Partial root-zone irrigation technique has also been studied to reduce water consumption with low yield loss. Alternate partial root-zone irrigation, in which half of the root system is irrigated normally, while the other half is exposed to drying soil, improves WUE, enhances root activity and can increase yield, when compared to conventional drip irrigation [119]. Bench sub-irrigation system reduces plant height, leaf area, total fresh and dry weight but as NS management is very simple with this technique, it could be used for closed cycle cultivation even in low-tech greenhouses [120] (Montesano et al., 2010). Water pillow irrigation is a new method with quite high irrigation water efficiency studied by [115]. The results in greenhouse tomato showed that irrigation water using water pillow was 52% less than in drip irrigation, the total yield and yield per plant were 17% higher, and no weed development was observed in water pillow in contrast with drip irrigation, and fruits had better physic-chemical characteristics (pH, titration acidity, °brix, total dry matter, and color values).

In soilless culture, the NS can be single-used in an open, non-circulating system or reused in closed, recirculating systems [107]. Non-circulating systems use NS supplied in each irrigation and account for runoff losses [107] while recirculating systems collect leftover NS, blend it with fresh NS, and recirculate it in subsequent irrigations [29, 120], increasing water and nutrient use efficiency, despite some yield reduction that can occur, induced by salinization in root zone [2, 107, 121]. One of the main problems associated with these systems is the potential spread of plant root pathogens and the accumulation of other chemical compounds [122]. Most growers use physical and chemical disinfection methods to treat the drain water and reduce the risk of spreading root diseases. Pasteurization, ultraviolet treatment or chlorine dosing are the most common treatments [29] but heat, sonication, application of copper and silver ions, active carbon absorption, hydrogen peroxide, as well as dissolution of ozone into bulk irrigation solutions are used as well. Ozone effectively reduces chemical contaminant and pathogen levels in greenhouse irrigation water [123].

High NS electrical conductivity enhances sensorial quality of greenhouse tomatoes as studies reveal that under this condition, fruits are more flavourful, redder, smaller, softer at touch, firmer in the mouth, crunchier and sourer, being usually preferred by consumers in sensorial tests [121]. The sensory quality is accompanied by higher sugar and acid content and aroma volatiles [124, 125]. Elevating the EC of the NS can thus be a simple way to improve tomato fruit quality [124].

In the Mediterranean, deficit irrigation and intensive cultivation are common practices which leads to salinization, poor irrigation, and poor soil quality. In a simulated Mediterranean greenhouse, under high salinity irrigation ECw = 3.5 dS m^{-1}), [126] concluded that early harvest and early termination of the season have no significant impact on tomato yield while alleviating the pressure over natural resources.

Inducing deficit irrigation can promote quality to some extent [127, 128], but scheduling could greatly affect the results [128]. Regulated deficit irrigation (RDI) is meant to ensure an optimal crop water status in most sensitive to water-stress phenological phases and restrict irrigation in the other development phases [129]. Several studies have been made, and the results clearly point out that WUE increase although the yield is reduced tomato [130]. In northwest China, [131] determined that the application of 1/3 or 2/3 of the normally full irrigation amount at flowering and fruit development stages, and no water stress in the other growth stages, can result in a good concession between yield and quality of tomato. On the contrary, [128] showed that, at the seedling stage, the application of 1/3 and 2/3 of full irrigation did not significantly impact water consumption, total yield and fruit quality, but the application of 1/3 of full irrigation at flowering and fruit development stage had a negative effect in total yield. The application of 1/3 or 2/3 of full irrigation at the fruit maturation stage also impacts negatively in total yield. However, despite the negative results in yield, fruit quality was significantly increased, concluding that tomato yield is sensitive to water deficit during the fruit development stage and fruit maturation stage, while fruit quality is mainly affected by water stress during the fruit maturation stage. The application of 2/3 full irrigation at fruit development stage had no significant effects on yield and quality. Some authors defend that RDI based on leaf water potential has the potential to improve the accuracy of irrigation scheduling, since it integrates the soil-water-plant relations, which could lead to important reductions on yield loss [130].

Last trends in greenhouse water use (and in agriculture in general), point to reclaimed water utilization of various sources, boosted by recent legislative changes in the UE, and, namely, in the south, by increasing water scarcity. It is expected that in the following years, the use of these water sources also sees an intensification to fight climate change consequences, although there are some challenges and concerns that need to be addressed to its widespread use. Israel is pioneer in reclaimed water use in agriculture, since water scarcity in the country is a hard reality, but even there, concerns are considered, and precautions are being taken to minimize and overcome health security risks [132].

Most studies approach the use of treated urban wastewater, but other alternatives are also considered as reuse rainfall, drainage from greenhouses and aquaponic water, with positive results in yield and quality but fewer details about pathogenic or compound accumulation in fruit [30, 133–135] which consists of vital information for assessing the viability of the use of these resources. As an example, in Portugal, the reuse of NS in greenhouses has been used for some producers and for some time now, but there was no evidence in literature about this practice in the country until 2018 with the announcement of TomatInov Project and the consecutive divulgation and demonstration actions [136] where the NS is reused in a semi-closed system and the demonstration, in 2020, of full closed system by New Growing Systems [137], which reveals that although many technologies are in use, there is a lot of room for improvement.

8. Conclusion and directions

The increasing challenges related to anticipated global climate change, increasing world population and demand for healthier products push the research and the application of more efficient production techniques. Reducing greenhouse energy consumption by increasing energy efficiency and by the application of renewable energy sources is imperative. Face water shortages by changing irrigation methods and pursuing water use efficiency as well as the use of reclaimed water, control environmental conditions, use tailored light only where necessary and sustainable substrate materials, breed cultivars that are more resistant to biotic and abiotic stresses, and the use of rootstock to induce resistance or vigor but generate non-GMO products alleviating biosafety concerns, are all pieces of an integrated strategy to achieve the goal of sustainable production.

Following worldwide research trends and policies, the use of reclaimed water or reuse of treated NS or wastewater from other activities is expected to be a great advance in the Mediterranean region in the next years, and, hopefully, becomes a generalized and safe practice in a region suffering with climate change and water shortage.

Acknowledgements

TomatInov Project PDR2020-101-032136 is promoted by PDR2020 and co-financed by FEADER under the Portugal 2020 initiative, Action 1, Operational groups.

Conflict of interest

The authors have no competing interests to declare.

Author details

Raquel Saraiva[1,2*], Igor Dias[1,3], José Grego[1] and Margarida Oliveira[1,2,4]

1 Escola Superior Agrária, UIIPS - Instituto Politécnico de Santarém, Santarém, Portugal

2 LEAF Linking Landscape, Environment, Agriculture and Food Research Center, Associated Laboratory TERRA, Instituto Superior de Agronomia, Universidade de Lisboa, Lisboa, Portugal

3 MED - Mediterranean Institute of Agriculture, Environment and Development and Global Change and Sustainability Institute, Associated Laboratory Change, Universidade de Évora, Portugal

4 CIEQV – Life Quality Research Centre, Rio Maior, Portugal

*Address all correspondence to: raquel.saraiva@esa.ipsantarem.pt

References

[1] United Nations. Goals for Sustainable Development. 2015. Available from: https://www.un.org/sustainabledevelopment/sustainable-consumption-production/

[2] Pardossi A, Tognoni F, Incrocci L. Mediterranean greenhouse technology. Chronica Horticulturae. 2004;**44**:28-34

[3] Mariz-Ponte N, Martins S, Gonçalves A, Correia CM, Ribeiro C, Santos MCDC. The potential use of the UV-A and UV-B to improve tomato quality and preference for consumers. Scientia Horticulturae. 2019;**246**:777-784. DOI: 10.1016/j.scienta.2018.11.058

[4] FAO. Countries by Commodity. 2023. Available from: https://www.fao.org/faostat/en/#rankings/countries_by_commodity

[5] CONFAGRI. Como vai mudar o mercado agrícola europeu até 2030?, 2018, available from: https://www.confagri.pt/vai-mudar-mercado-agricola-europeu-ate-2030/

[6] Paulo Maria. Personal communication, Hortomaria, Torres vedras. 2019

[7] Gruda N, Tanny J. Protected crops – Recent advances, innovative technologies and future challenges. Acta Horticulturae. 2015;**1107**:271-278. DOI: 10.17660/ActaHortic.2015.1107.37

[8] FAO. Plant Production and Protection Paper 230- Good Agricultural Practices for greenhouse vegetable production in the South East European countries. 2017. ISSN 2070-2515

[9] Muñoz P, Antón A, Nuñez M, Paranjpe A, Ariño J, Castells X, et al. Comparing the environmental impacts of greenhouse versus open-field tomato production in the Mediterranean region. GreenSys 2007: High Technology for Greenhouse System Management. Italy: Naples. 3-6 Oct 2007;**1**:1591-1596

[10] Baptista F, Murcho D, Silva L. Assessment of Energy Consumption in Greenhouse Tomato Production in the West Region of Portugal. Portugal: Greensys; 2015

[11] Ferreira VS. Avaliação das condições climáticas em dois tipos de estufa e sua influência na produtividade e nos custos de produção do tomate, na região do Oeste, Master Thesys. ISA: Ulisboa; 2017

[12] Salazar R, Rojano AAL. New technologies in tomato greenhouse production. In: Proceedings International Conference of Agricultural Engineering. Zurich: EURAGENG; 2014

[13] Hochmuth GJ. Production of Greenhouse Tomatoes - Florida Greenhouse Vegetable Production Handbook. 2018. Vol. 3. Available from: http://edis.ifas.ufl.edu/pdffiles/CV/CV26600.pdf

[14] European Scientists. Position Paper on the ECJ Ruling on CRISPR. 2018. Available from: www.vib.be/en/news/Documents/

[15] European Scientists. Letter to President Juncker on Genome Editing. 2019. Available from: http://www.vib.be

[16] Schwarz D, Öztekin GB, Tüzel Y, Brückner B, Krumbein A. Rootstocks can enhance tomato growth and quality characteristics at low potassium supply. Scientia Horticulturae. 2013;**149**:70-79. DOI: 10.1016/j.scienta.2012.06.013

[17] Bai M, Chen W, Xie B, Yang G. A novel strategy to enhance resistance to cucumber mosaic virus in tomato by grafting to transgenic rootstocks. Journal of Integrative Agriculture. 2016;**15**(9):2040-2048. DOI: 10.1016/S2095-3119(16)61330-8

[18] Asins MJ, Albacete A, Martinez-Andujar C, Pérez-Alfocea F, Dieleman JA. Genetic analysis of rootstock-mediated nitrogen (N) uptake and root-to-shoot signalling at contrasting N availabilities in tomato. Plant Science. 2017;**263**:94-106. DOI: 10.1016/j.plantsci.2017.06.012

[19] Cantero-Navarro E, Romero-Aranda R, Fernández-Muñoz R, Martínez-Andújar C, Albacete A. Improving agronomic water use efficiency in tomato by rootstock-mediated hormonal regulation of leaf biomass. Plant Science. 2016;**251**:90-100. DOI: 10.1016/j.plantsci.2016.03.001

[20] Hanna HY. Influence of cultivar, growing media, and cluster pruning on greenhouse tomato yield and fruit quality. HortThechnology. 2009;**19**(2):395-399. DOI: 10.21273/HORTSCI.19.2.395

[21] Zhang X, You S, Tian Y, Li J. Comparison of plastic film, biodegradable paper and bio-based film mulching for summer tomato production: Soil properties, plant growth, fruit yield and fruit quality. Scientia Horticulturae. 2019;**249**:38-48. DOI: 10.1016/j.scienta.2019.01.037

[22] Xiaohui F, Kai G, Ce Y, Jinsong L, Huanyu C, Xiaojing L. Growth and fruit production of tomato grafted onto wolfberry (Lycium chinense) rootstock in saline soil. Scientia Horticulturae. 2019;**255**:298-305. DOI: 10.1016/j.scienta.2019.05.028

[23] Bello TT, Fawole B, Claudius-Cole A. Management of Root-Knot Nematodes (Meloidogyne Spp) on tomato using antagonistic plants. Journal of Biology, Agriculture and Healthcare. 2014;**4**(24):97-101

[24] Gómez-Tenorio MA, Tello JC, Zanón MJ, de Cara M. Soil disinfestation with dimethyl disulfide (DMDS) to control Meloidogyne and Fusarium oxysporum f. sp. radicis-lycopersici in a tomato greenhouse. Crop Protection. 2018;**112**:133-140. DOI: 10.1016/j.cropro.2018.05.023

[25] Chellemi DO, Rosskopf EN. Yield potential and soil quality under alternative crop production practices for fresh market pepper. Renewable Agriculture and Food Systems. 2004;**19**:168-175. DOI: 10.1079/RAFS200479

[26] Miller SA. The Ohio State University. Managing Diseases of Organic Tomatoes in Greenhouses and High Tunnels. 2018. Available from: https://eorganic.org/node/3069

[27] Matthiessen JN, Kirkegaard JA. Biofumigation and enhanced biodegradation: Opportunity and challenge in soilborne pest and disease management. Critical Reviews in Plant Sciences. 2006;**25**:235-265. DOI: 10.1080/07352680600611543

[28] Mauro RP, Lo Monaco A, Lombardo S, Restuccia A, Mauromicale G. Eradication of Orobanche/Phelipanche spp. seedbank by soil solarization and organic supplementation. Scientia Horticulturae. 2015;**193**:62-68. DOI: 10.1016/j.scienta.2015.06.038

[29] Selina P, Bledsoe ME. U.S. Greenhouse/Hothouse Hydroponic Tomato Timeline. Raleigh, NC 27606: Village Farms, L.P; 2002. Available from: https://ipmdata.ipmcenters.org

[30] Lopez-Galvez F, Allende A, Pedrero-Salcedo F, Alarcon JJ, Gil MI. Safety assessment of greenhouse hydroponic tomatoes irrigated with reclaimed and surface water. International Journal of Food Microbiology. 2014;**191**:97-102. DOI: 10.1016/j.ijfoodmicro.2014.09.004

[31] Papadopoulos AP. Growing Greenhouse Tomatoes in Soil and in Soilless Media. Ontario, Canada: Publ. 1865/E, Communi. Branch, Agriculture and Agri-food; 1991

[32] Dannehl D, Suhl J, Ulrichs C, Schmidt U. Evaluation of substitutes for rock wool as growing substrate for hydroponic tomato production. Journal of Applied Botany and Food Quality. 2015;**88**:68-77. DOI: 10.5073/JABFQ.2015.088.010

[33] Kaur H. Development of Hydroponic System for Greenhouse Tomato, Master Thesis in Soil and Water Engineering. Punjab: Punjab Agricultural University; 2016

[34] Valenzano V, Parente A, Serio F, Santamaria P. Effect of growing system and cultivar on yield and water-use efficiency of greenhouse-grown tomato. Journal of Horticultural Science & Biotechnology. 2008;**83**(1):71-75

[35] Almaghrabi OA, Massoud SI, Abdelmoneim TS. Influence of inoculation with plant growth promoting rhizobacteria (PGPR) on tomato plant growth and nematode reproduction under greenhouse conditions. Saudi Journal of Biological Sciences. 2013;**20**(1):57-61. DOI: 10.1016/j.sjbs.2012.10.004

[36] Punja ZK, Rodriguez G, Tirajoh A. Effects of Bacillus subtilis strain QST 713 and storage temperatures on post-harvest disease development on greenhouse tomatoes. Crop Protection. 2016;**84**:98-104. DOI: 10.1016/j.cropro.2016.02.011

[37] Gravel V, Antoun H, Tweddell RJ. Growth stimulation and fruit yield improvement of greenhouse tomato plants by inoculation with pseudomonas putida or Trichoderma atroviride: Possible role of indole acetic acid (IAA). Soil Biology and Biochemistry. 2007;**39**(8):1968-1977. DOI: 10.1016/j.soilbio.2007.02.015

[38] Yim W, Seshadri S, Kim K, Lee G, Sa T. Ethylene emission and PR protein synthesis in ACC deaminase producing Methylobacterium spp. inoculated tomato plants (Lycopersicon esculentum mill.) challenged with Ralstonia solanacearum under greenhouse conditions. Plant Physiology and Biochemistry. 2013;**67**:95-104. DOI: 10.1016/j.plaphy.2013.03.002

[39] Delaplane KS, Mayer DR, Mayer DF. Crop Pollination by Bees. Wallingford: CABI Publishing; 2000. p. 352. ISBN: 085199783X, 9780851997834

[40] Almeida L, Amaro F, Figueiredo E, Godinho MC, Mexia A, Nunes AP. Melhoria da produção hortícola em estufa na região oeste. 1999. Available from: http://www.cothn.pt

[41] Borges JCMC. Bombus terrestris e Bombus spp; Captura, Identificação das Espécies, Subespécies e Castas, Ciclo de Vida e Dinâmica Populacional das Populações Selvagens, Master Thesys. Porto: FC da Universidade do Porto; 2018

[42] Kapongo JP, Shipp L, Kevan P, Sutton JC. Co-vectoring of Beauveria bassiana and Clonostachys rosea by bumble bees (Bombus impatiens) for control of insect pests and suppression of grey mould in greenhouse tomato and sweet pepper. Biological Control. 2008;**46**(3):508-514. DOI: 10.1016/j.biocontrol.2008.05.008

[43] Seabra SG, Silva SE, Nunes VL, Sousa VC, Martins J, Marabuto E, et al. Genomic signatures of introgression between commercial and native bumblebees, Bombus terrestris, in western Iberian Peninsula – Implications for conservation and trade regulation. Evolutionary Applications. 2018;**12**(4):679-691. DOI: 10.1111/eva.12732

[44] Jarquín-Enríquez L, Mercado-Silva EM, Maldonado JL, Lopez-Baltazar J. Lycopene content and color index of tomatoes are affected by the greenhouse cover. Scientia Horticulturae. 2013;**155**:43-48. DOI: 10.1016/j.scienta.2013.03.004

[45] Kavga A, Souliotis M, Koumoulose EP, Fokaidesf PA, Charitidise CA. Environmental and nanomechanical testing of an alternative polymer nanocomposite greenhouse covering material. Solar Energy. 2018;**159**:1-9. DOI: 10.1016/j.solener.2017.10.073

[46] Geoola F, Kashti Y, Levi A, Brickman R. Quality evaluation of antidrop properties of greenhouse cladding materials. Polymer Testing. 2004;**23**:755-761. DOI: 10.1016/j.polymertesting.2004.04.006

[47] Tantau HJ, Hinken J, von Elsner B, Max JFJ, Ulbrich A, Schurr U, et al. Solar transmittance of greenhouse covering materials. Acta Horticulturae. 2012;**956**:441-448. DOI: 10.17660/ActaHortic.2012.956.51

[48] Emekli NY, Büyüktaş K, Başçetinçelik A. Changes of the light transmittance of the LDPE films during the service life for greenhouse application. Journal of Building Engineering. 2016;**6**:126-132. DOI: 10.1016/j.jobe.2016.02.013

[49] Alsadon A, Al-Helal I, Ibrahim A, Abdel-Ghany A, Al-Zaharani S, Ashour T. The effects of plastic greenhouse covering on cucumber (Cucumis sativus L.) growth. Ecological Engineering. 2015;**87**:305-312. DOI: 10.1016/j.ecoleng.2015.12.005

[50] Max JFJ, Schurr U, Tantau HJ, Mutwiwa UN, Hofmann T, Ulbrich A. Greenhouse Cover Technology in Horticultural Reviews. In: Jules J, editor. Wiley- Blackwell. 2012;**40**. DOI: 10.1002/9781118351871.ch7. ISBN: 9781118345832

[51] Aldaftari HA, Okajima J, Komiya A, Maruyama S. Radiative control through greenhouse covering materials using pigmented coatings. Journal of Quantitative Spectroscopy and Radiative Transfer. 2019;**231**:29-36. DOI: 10.1016/j.jqsrt.2019.04.009

[52] Xie X, Liu YJ, Hao JJ, Ju L, Du WC, Yang HW. Feasibility study of a new solar greenhouse covering material. Journal of Quantitative Spectroscopy and Radiative. 2019;**224**:37-43. DOI: 10.1016/j.jqsrt.2018.11.004

[53] Al-Mahdouri A, Gonome H, Okajima J, Maruyama S. Theoretical and experimental study of solar thermal performance of different greenhouse cladding materials. Solar Energy. 2014;**107**:314-327. DOI: 10.1016/j.solener.2014.05.006

[54] Castilla N, Baeza E. Greenhouse Site Selection, Chap.2. Rome: FAO Paper 217, Food and Agriculture Organization of the United Nations; 2013

[55] Kwon JK, Khoshimkhujaev B, Lee JH, Yu IH, Park KS, Choi HG. Growth and yield of tomato and cucumber plants in polycarbonate or glass greenhouses. Korean Journal of Horticultural Science and Technology. 2017;**35**(1):79-87. DOI: 10.12972/kjhst.20170009

[56] Živanović B, Vidović M, Komić SM, Jovanović L, Kolarž P, Morina F, et al. Contents of phenolics and carotenoids in tomato grown under polytunnels with different UV-transmission rates. Turkish Journal of Agriculture and Forestry. 2017;**41**(2):113-120. DOI: 10.3906/tar-1612-56

[57] Murakami K, Fukuok N, Noto S. Improvement of greenhouse microenvironment and sweetness of melon (Cucumis melo L.) fruits by greenhouse shading with a new kind of near-infrared ray-cutting net in mid-summer. Scientia Horticulturae. 2018;**218**:1-7. DOI: 10.1016/j.scienta.2017.02.011

[58] Kittas C, Katsoulas N, Rigakis N, Bartzanas T, Kitta E. Effects on microclimate, crop production and quality of a tomato crop grown under shade nets. The Journal of Horticultural Science and Biotechnology. 2012;**87**:7-12. DOI: 10.1080/14620316.2012.11512822

[59] Smith GB, Gentle A, Swift PD, Earp A, Mronga N. Coloured paints based on iron oxide and silicon oxide coated flakes of aluminium as the pigment, for energy efficient paint: Optical and thermal experiments. Solar Energy Materials & Solar Cells. 2003;**79**(2):179-197. DOI: 10.1016/S0927-0248, (02)00410-5

[60] Seven SA, Tastan ÖF, Tas CE, Ünal H, Ince İA, Menceloglu YZ. Insecticide-releasing LLDPE films as greenhouse cover materials. Materials Today Communications. 2019;**19**:170-176. DOI: 10.1016/j.mtcomm.2019.01.015

[61] Jiang C, Johkan M, Hohjo M, Tsukagoshi S, Maruo T. Photosynthesis, plant growth, and fruit production of single-truss tomato improves with supplemental lighting provided from underneath or within the inner canopy. Scientia Horticulturae. 2017;**222**:221-229

[62] Kowalczyk K, Gajc-Wolska J, Metera A, Mazur K, Radzanowska J, Szatkowski M. Effect of supplemental lighting on the quality of tomato fruit (*Solanum lycopersicum* M L.) in autumn-winter cultivation. Acta Horticulturae. 2012;**956**:395-401. DOI: 10.17660/ActaHortic.2012.956.46

[63] Hao X, Little C, Zheng JM, Cao R. Far-red LEDs improve fruit production in greenhouse tomato grown under high-pressure sodium lighting. Acta Horticulturae. 2016;**1134**:95-102. DOI: 10.17660/ActaHortic.2016.1134.13

[64] Zhang YT, Zhang YQ, Yang QC, Li T. Overhead supplemental far-red light stimulates tomato growth under intra-canopy lighting with LEDs. Journal of Integrative Agriculture. 2019;**18**(1):62-69. DOI: 10.1016/S2095-3119(18)62130-6

[65] Fanwoua J, Vercambre G, Buck-Sorlin G, Dieleman JA, Génard M. Supplemental LED lighting affects the dynamics of tomato fruit growth and composition. Scientia Horticulturae. 2019;**256**:108571. DOI: 10.1016/j.scienta.2019.108571

[66] Mitchell C, Dzakovich M, Gomez C, Lopez R, Burr J, Hernández R, et al. Light-emitting diodes in horticulture. Horticultural Reviews. 2015;**43**:1-88. DOI: 10.1002/9781119107781.ch01

[67] Dzakovich MP, Gómez C, Mitchell CA. Tomatoes grown with light-emitting diodes or high-pressure sodium supplemental lights have similar fruit-quality attributes. HortScience. 2015;**50**:1498-1502. DOI: 10.21273/HORTSCI.50.10.1498

[68] Deram P. Light-Emitting-Diode (LED) Lighting for Greenhouse Tomato Production, Master thesis, McGill University; Montreal, 2013

[69] Christie JM. Phototropin blue-light receptors. Annual Review of Plant Biology. 2007;**58**:21-45. DOI: 10.1146/annurev.arplant.58.032806.103951

[70] O'Carrigan A, Hinde E, Lu N, Xu XQ, Duan H, Huang G, et al. Effects of light irradiance on stomatal regulation and growth of tomato. Environmental and Experimental Botany. 2014;**98**:65-73. DOI: 10.1016/j.envexpbot.2013.10.007

[71] Smith HL, McAusland L, Murchie EH. Don't ignore the green light: Exploring diverse roles in plant processes. Journal of Experimental Botany. 2017;**68**:2099-2110. DOI: 10.1093/jxb/erx098

[72] Lin KH, Huang MY, Huang WD, Hsu MH, Yang ZW, Yang CM. The effects of red, blue, and white light-emitting diodes on the growth, development, and edible quality of hydroponically grown lettuce (Lactuca sativa L. var. capitata). Scientia Horticulturae. 2013;**150**:86-91. DOI: 10.1016/j.scienta.2012.10.002

[73] Kaiser E, Weerheim K, Schipper R, Dieleman JA. Partial replacement of red and blue by green light increases biomass and yield in tomato. Scientia Horticulturae. 2019;**249**:271-279. DOI: 10.1016/j.scienta.2019.02.005

[74] Schouten RE, Woltering EJ, Tijskens LMM. Sugar and acid interconversion in tomato fruits based on biopsy sampling of locule gel and pericarp tissue. Postharvest Biology and Technology. 2016;**111**:83-92. DOI: 10.1016/j.postharvbio.2015.07.032

[75] Kim HJ, Yang T, Choi S, Wang YJ, Lin MY, Liceaga AM. Supplemental intracanopy far-red radiation to red LED light improves fruit quality attributes of greenhouse tomatoes. Scientia Horticulturae. 2020;**261**:108985. DOI: 10.1016/j.scienta.2019.108985

[76] Gómez C, Mitchell CA. In search of an optimized supplemental lighting spectrum for greenhouse tomato production with intracanopy lighting. Acta Horticulturae. 2016;**1134**:57-62. DOI: 10.17660/ActaHortic.2016.1134.8

[77] Li Q, Liu Y, Tian S, Liang Z, Li S, Li Y, et al. Effect of supplemental lighting on water transport, photosynthetic carbon gain and water use efficiency in greenhouse tomato. Scientia Horticulturae. 2019;**256**:108630. DOI: 10.1016/j.scienta.2019.108630

[78] Lu N, Maruo T, Johkan M, Hohjo M, Tsukagoshi S, Ito Y, et al. Effects of supplemental lighting with light-emitting diodes (LEDs) on tomato yield and quality of single-truss tomato plants grown at high planting density environ. Control Biology. 2012;**50**(1):63-74. DOI: 10.2525/ecb.50.63

[79] Grozeff GEG, Senn ME, Alegre ML, Chaves AR, Bartoli CG. Nocturnal low irradiance pulses improve fruit yield and lycopene concentration in tomato. Scientia Horticulturae. 2016;**203**:47-52. DOI: 10.1016/j.scienta.2016.03.001

[80] De Pascale S, Maggio A. Sustainable protected cultivation at Mediterranean climate, perspectives and challenges. Acta Horticulturae. 2004;**691**:29-42. DOI: 10.17660/ActaHortic.2005.691.1

[81] Shen Y, Wei R, Xu L. Energy consumption prediction of a greenhouse and optimization of daily average temperature. Energies. 2018;**11**:65. DOI: 10.3390/en11010065

[82] Baptista FJ, Briassoulis D, Stanghellini C, Silva LL, Balafoutis AT, Meyer-Aurich A, et al. Energy efficiency In tomato greenhouse production - a

preliminary study. Acta Horticulturae. 2014;**1037**:179-185. DOI: 10.17660/ActaHortic.2014.1037.18

[83] Gourdo L, Fatnassi H, Tiskatine R, Wifaya A, Demrati H, Aharoune A, et al. Solar energy storing rock-bed to heat an agricultural greenhouse. Energy. 2019;**169**:206-212. DOI: 10.1016/j.energy.2018.12.036

[84] Ntinas GK, Koukounaras A, Kotsopoulos T. Effect of energy saving solar sleeves on characteristics of hydroponic tomatoes grown in a greenhouse. Scientia Horticulturae. 2015;**194**:126-133. DOI: 10.1016/j.scienta.2015.08.013

[85] Bazgaou A, Fatnassi H, Bouharroud R, Ezzaeri K, Gourdo L, Wifaya A, et al. Effect of active solar heating system on microclimate, development, yield and fruit quality in greenhouse tomato production. Renewable Energy. 2021;**165**(Part 1):237-250. DOI: 10.1016/j.renene.2020.11.007

[86] Gourdo L, Fatnassi H, Bouharroud R, Ezzaeri K, Bazgaou A, Wifaya A, et al. Heating canarian greenhouse with a passive solar water–sleeve system: Effect on microclimate and tomato crop yield. Solar Energy. 2019;**188**:1349-1359. DOI: 10.1016/j.solener.2019.07.004

[87] Mobtaker HG, Ajabshirchi Y, Ranjbar F, Matloobi M. Solar energy conservation in greenhouse: Thermal analysis and experimental validation. Renewable Energy. 2016;**96**:509-519. DOI: 10.1016/j.renene.2016.04.079

[88] Hassanien RHE, Li M, Yin F. The integration of semi-transparent photovoltaics on greenhouse roof for energy and plant production. Renewable Energy. 2018;**121**:377-388. DOI: 10.1016/j.renene.2018.01.044

[89] Callejón-Ferre AJ, Velázquez-Martí B, López-Martínez JA, Manzano-Agugliaro F. Greenhouse crop residues: Energy potential and models for the prediction of their higher heating value. Renewable and Sustainable Energy Reviews. 2011;**15**:948-955. DOI: 10.1016/j.rser.2010.11.012

[90] Kläring HP, Klopotek Y, Krumbein A, Schwarz D. The effect of reducing the heating set point on the photosynthesis, growth, yield and fruit quality in greenhouse tomato production. Agricultural and Forest Meteorology. 2015;**214**:178-188. DOI: 10.1016/j.agrformet.2015.08.250

[91] Baptista F. Modelling the Climate in Unheated Tomato Greenhouses and Predicting Botrytis Cinerea Infection. Doctoral Thesis. Évora: Évora University; 2007

[92] Kumar KS, Jha MK, Tiwari KN, Singh A. Modeling and evaluation of greenhouse for floriculture in subtropics. Energy and Buildings. 2010;**42**(7):1075-1083. DOI: 10.1016/j.enbuild.2010.01.021

[93] Qian T. Crop Growth and Development in Closed and Semi-Closed Greenhouses. Wageningen, NL: Wageningen University; 2017. p. 112. PhD thesis

[94] Zhang Z, Gates R, Zhirong Z, Xiaohui H. Evaluation of ventilation performance and energy efficiency of greenhouse fans. International Journal of Agricultural and Biological Engineering. 2015;**8**:103. DOI: 10.3965/j.ijabe.20150801.014

[95] Hou H, Chen H, Cai H, Yang F, Wang F. CO_2 and N_2O emissions from Lou soils of greenhouse tomato fields under aerated irrigation. Atmospheric Environment. 2016;**132**:69-76. DOI: 10.1016/j.atmosenv.2016.02.027

[96] Li Y, Niu W, Dyck M, Wang J, Zou X. Yields and nutritional of greenhouse tomato in response to different soil aeration Vol. at two depths of subsurface drip irrigation. Scientific Reports. 2016;**6**:39307

[97] Chen H, Hou HJ, Wang XY, Zhu Y, Cai H. The effects of aeration and irrigation regimes on soil CO_2 and N_2O emissions in a greenhouse tomato production system. Journal of Integrative Agriculture. 2018;**17**(2):449-460. DOI: 10.1016/S2095-3119(17)61761-1

[98] Du YD, Gu XB, Wang JW, Niu WQ. Yield and gas exchange of greenhouse tomato at different nitrogen levels under aerated irrigation. Science of the Total Environment. 2019;**668**(10):1156-1164. DOI: 10.1016/j.scitotenv.2019.03.098

[99] Zhang Z, Liu L, Zhang M, Zhang Y, Wang Q. Effect of carbon dioxide enrichment on health-promoting compounds and organoleptic properties of tomato fruits grown in greenhouse. Food Chemistry. 2014;**153**:157-163. DOI: 10.1016/j.foodchem.2013.12.052

[100] Jin C, Du S, Wang Y, Condon J, Lin X, Zhang Y. Carbon dioxide enrichment by composting in greenhouses and its effect on vegetable production. Journal of Plant Nutrition and Soil Science. 2009;**172**:418-424. DOI: 10.1002/jpln.200700220

[101] Marchi B, Zanoni S, Pasetti M. Industrial Symbiosis for greener horticulture practices: The CO2 enrichment from energy intensive industrial processes. Procedia CIRP. 2018;**69**:562-567. DOI: 10.1016/j.procir.2017.11.117[

[102] Stanghellini C, Kempkes F. Steering of fogging: Control of humidity, temperature or transpiration? Acta Horticulturae. 2008;**797**:61-67. DOI: 10.17660/ActaHortic.2008.797.6

[103] Leyva R, Constán-Aguilar C, Sánchez-Rodríguez E, Romero-Gámez M, Soriano T. Cooling systems in screenhouses: Effect on microclimate, productivity and plant response in a tomato crop. Biosystems Engineering. 2015;**129**:100-111. DOI: 10.1016/j.biosystemseng.2014.09.018

[104] Abdel-Ghany AM, Kozai T. Cooling efficiency of fogging Systems for Greenhouses. Biosystems Engineering. 2006;**94**:97-109. DOI: 10.1016/j.biosystemseng.2006.02.008

[105] Kaneda Y, Ibayashi H, Oishi N, Mineno H. Greenhouse environmental control system based on SW-SVR. Procedia Computer Science. 2015;**60**:860-869. DOI: 10.1016/j.procs.2015.08.249

[106] Kurose T. Tomato king Kagome leads with high-tech greenhouses. In Nikkei Asian Review. 2016. Available from: https://asia.nikkei.com/Business/Biotechnology/

[107] Meric MK, Tuzel IH, Tuzel Y, Oztekin GB. Effects of nutrition systems and irrigation programs on tomato in soilless culture. Agricultural Water Management. 2011;**99**(1):19-25. DOI: 10.1016/j.agwat.2011.08.004

[108] Massa D, Incrocci L, Maggini R, Bibbiani C, Pardossi A. Simulation of crop water and mineral relations in greenhouse soilless culture. Environmental Modelling & Software. 2011;**26**(6):711-722. DOI: 10.1016/j.envsoft.2011.01.004

[109] Kondo N, Yata K, Iida M, Shiigi T, Omori H. Development of an end-effector for a tomato cluster harvesting robot. Engineering in Agriculture, Environment and Food.

2014;**3**(1):20-24. DOI: 10.1016/S1881-8366(10)80007-2

[110] Metomotion. GRoW - Greenhouse Robotic Worker. 2018. Available from: https://metomotion.com/

[111] Li Z, Miao F, Yang Z, Chai P, Yang S. Factors affecting human hand grasp type in tomato fruit-picking: A statistical investigation for ergonomic development of harvesting robot. Computers and Electronics in Agriculture. 2019;**157**:90-97. DOI: 10.1016/j.compag.2018.12.047

[112] Li Z, Li P, Liu J. Physical and mechanical properties of tomato fruits as related to robot's harvesting. Journal of Food Engineering. 2011;**103**(2):170-178. DOI: 10.1016/j.jfoodeng.2010.10.013

[113] García-Luna F, Morales-Díaz A. Towards an artificial vision-robotic system for tomato identification. IFAC-PapersOnLine. 2016;**49**(16):365-370

[114] Rincón VJ, Grella M, Marucco P, Alcatrão LE, Sanchez-Hermosilla J, Balsari P. Spray performance assessment of a remote-controlled vehicle prototype for pesticide application in greenhouse tomato crops. Science of the Total Environment. 2020;**726**:138509. DOI: 10.1016/j.scitotenv.2020.138509

[115] Gerçek S, Demirkaya M, Işik D. Water pillow irrigation versus drip irrigation with regard to growth and yield of tomato grown under greenhouse conditions in a semi-arid region. Agricultural Water Management. 2017;**180**:172-177. DOI: 10.1016/j.agwat.2016.11.012

[116] Li Y, Wang L, Xue X, Guo W, Chen F. Comparison of drip fertigation and negative pressure fertigation on soil water dynamics and water use efficiency of greenhouse tomato grown in the North China plain. Agricultural Water Management. 2017;**184**:1-8. DOI: 10.1016/j.agwat.2016.12.018

[117] Li X, Xie YG, Wang GD, Xue XZ. Growth and physiological characteristics of eggplant under negative hydraulic head irrigation in greenhouse. Natural Sciences Education. 2016;**44**:163-169. DOI: 10.3389/fpls.2015.00262

[118] Li S, Xue XZ, Guo WS, Li X, Chen F. Effects of water supply tension on yield and water use efficiency of greenhouse cucumber. Scientia Agricultura Sinica. 2010;**43**:337-345. DOI: 10.3864/j.issn.0578-1752.2010.02.014

[119] Yang L, Qu H, Zhang Y, Li F. Effects of partial root-zone irrigation on physiology, fruit yield and quality and water use efficiency of tomato under different calcium levels. Agricultural Water Management. 2012;**104**:89-94. DOI: 10.1016/j.agwat.2011.12.001

[120] Montesano F, Parente A, Santamaria P. Closed cycle subirrigation with low concentration nutrient solution can be used for soilless tomato production in saline conditions. Scientia Horticulturae. 2010;**124**:338-344. DOI: 10.1016/j.scienta.2010.01.017

[121] Cliff MA, Li JB, Toivonen PMA, Ehret DL. Effects of nutrient solution electrical conductivity on the compositional and sensory characteristics of greenhouse tomato fruit. Postharvest Biology and Technology. 2012;**74**:132-140. DOI: 10.1016/j.postharvbio.2011.12.007

[122] Rosberga AK, Gruyerb N, Hultberga M, Wohankac W, Alsanius BW. Monitoring rhizosphere microbial communities in healthy and Pythium ultimum inoculated tomato plants in soilless growing systems.

Scientia Horticulturae. 2014;**173**(27):106-113. DOI: 10.1016/j.scienta.2014.04.036

[123] Dannehl D, Schuch I, Gao Y, Cordiner S, Schmidt U. Hypochlorite application for controlling algae biofilm formation, microorganisms and tomato production in recirculating systems. Gesunde Pflanz. 2015;**67**:191-199. DOI: 10.1007/s10343-015-0351-3

[124] Dorais M, Dorval R, Demers DA, Micevic D, Turcotte G, Hao X, et al. Improving tomato fruit quality by increasing salinity: Effects on ion uptake, growth and yield. Acta Horticulturae. 2000;**511**:185-195. DOI: 10.17660/ActaHortic.2000.511.21

[125] Fabre MD, Jeannequin B. Effect of the salinity on the organoleptic quality and yield of early-grown soilless grafted tomatoes under heated glasshouses in the south of France. Cahiers Agricultures. 2011;**20**:266-273

[126] Daliakopoulos N, Apostolakis A, Wagner K, Deligianni A, Tsanis IK. Effectiveness of Trichoderma harzianum in soil and yield conservation of tomato crops under saline irrigation. Catena. 2019;**175**:144-153. DOI: 10.1016/j.catena.2018.12.009

[127] Favati F, Lovelli S, Galgano F, Miccolis V, Tommaso T, Candido V. Processing tomato quality as affected by irrigation scheduling. Scientia Horticulturae. 2009;**122**:562-571. DOI: 10.1016/j.scienta.2009.06.026

[128] Chen J, Kang S, Du T, Qiu R, Guo P, Chen R. Quantitative response of greenhouse tomato yield and quality to water deficit at different growth stages. Agricultural Water Management. 2013;**129**:152-162. DOI: 10.1016/j.agwat.2013.07.011

[129] Galindo A, Collado-González J, Griñán I, Corell M, Centeno A, Palomo MJ, et al. Deficit irrigation and emerging fruit crops as a strategy to save water in Mediterranean semiarid agrosystems. Agricultural Water Management. 2018;**202**:311-324. DOI: 10.1016/j.agwat.2017.08.015

[130] Coyago-Cruz E, Meléndez-Martínez AJ, Moriana A, Girón IF, Corell M. Yield response to regulated deficit irrigation of greenhouse cherry tomatoes. Agriculture Water Management. 2019;**213**:212-221. DOI: 10.1016/j.agwat.2018.10.020

[131] Wang F, Kang S, Du T, Li F, Qiu R. Determination of comprehensive quality index for tomato and its response to different irrigation treatments. Agricultural Water Management. 2011;**98**(8):1228-1238. DOI: 10.1016/j.agwat.2011.03.004

[132] Ministry of Health, state of Israel. Irrigation with Treated Wastewater. 2021. Available from: www.health.gov.il/English/Topics/EnviroHealth/Reclaimed_Water/

[133] Delaide B, Teerlinck S, Decombel A, Bleyaert P. Effect of wastewater from a pikeperch (Sander lucioperca L.) recirculated aquaculture system on hydroponic tomato production and quality. Agricultural Water Management. 2019;**226**:105814. DOI: 10.1016/j.agwat.2019.105814

[134] Magwaza ST, Magwaza LS, Odindo AO, Mditshwa A, Buckley C. Partially treated domestic wastewater as a nutrient source for tomatoes (Lycopersicum solanum) grown in a hydroponic system: Effect on nutrient absorption and yield. Heliyon. 2020;**6**:e05745. DOI: 10.1016/j.heliyon.2020.e05745

[135] Magwaza ST, Magwaza LS, Odindo AO, Mditshwa A, Buckley C.

Evaluating the feasibility of human excreta-derived material for the production of hydroponically grown tomato plants - part II: Growth and yield. Agricultural Water Management. 2020;**234**:106115. DOI: 10.1016/j.agwat.2020.106115

[136] Tomatinov. Acções. 2018. Available from: https://tomatinov.wordpress.com/

[137] Agrotec. Terraguar apresenta sistema inovador em Portugal para a produção de tomate em estufa, em sistema fechado. 2020. Available from: http://www.agrotec.pt

Chapter 5

Improving Tomato Productivity for Changing Climatic and Environmental Stress Conditions

Jithesh Mundaya Narayanan, Vishwini Viswanathan, Tulasi Tirumalai Ramanujam and Kruthika Nagendra Rao

Abstract

Tomato (*Lycopersicon esculentum* Mill.) growth, cultivation and its productivity are adversely influenced by severe environmental pressures. Several environmental abiotic factors that are limited not only to drought, salinity, temperature and heat but also to mechanical stress affect tomato yield and productivity. Salinity is a persistent problem throughout the world that affects soil properties. Further, tomato productivity due to salinity stress is affected at all stages of plant development. Seed priming, a method to alleviate salinity stress is an effective technique that can improve performance and growth. It is a method that permits controlled hydration of seeds thereby, maintaining metabolic activity, without allowing the protrusion of the radicle. Mechanical conditioning, a term applied to plant stimulation by tactile stimuli through various methods like touching, brushing, or rubbing the plant material, is another environmentally friendly and simple method to regulate plant growth and also stress tolerance. Therefore, the mechanical conditioning practice primes plants for enhanced plant growth and also allows plants to defend against an impending stress factor. These two methods can be developed into successful production practices. In this chapter, we summarize current knowledge of seed priming and mechanical conditioning for plant growth, cross-tolerance and plant productivity improvement.

Keywords: tomato, abiotic stress factors, salinity, seed priming, mechanical conditioning

1. Introduction

Plants are continuously exposed to a multitude of external stress factors and, are also influenced by the ever-changing environment. These stressful environmental conditions affects both plant growth, development, strength and also productivity [1]. Salinity, drought, temperature fluctuations, light intensity are some examples of stress conditions that act alone or in combination and cause wide-spread loss to plant productivity [2]. Apart from these factors, plants are also exposed to various forms of mechanical stress including wind, rain, pathogens, animals and even plants in nature.

Such forms of mechanical stress can result in adaptive responses, that are either thigmotropic (response to touch) or discreet (wounding) responses. These adaptive responses can be applied for plant growth, cross-stress tolerance and nutritional improvement [3].

One third of irrigated land has already deteriorated by salinity stress and, it is anticipated that by 2050, more than half (50%) of the world's cultivated land would be affected by soil salinity [4]. In the last few decades, engineering salinity tolerant crops through classical breeding, transgenics have gained considerable attention [5] and recently, genome editing has also gained significant attention [6]. However, due to the complex nature of the salt tolerance trait, these approaches have resulted in limited success [7].

1.1 Salinization

Soil salinization results in deleterious consequences in plants that are primarily caused by two mechanisms – osmotic stress and ion toxicity stress. The osmotic effect is short-term and occurs due to the uptake of Na^+ and Cl^- ions thereby, reducing the osmotic potential in plants. Ionic stress effects in plants are due to the high concentrations of Na^+ and Cl^- that invariably infiltrate the roots, decreases Na^+/K^+ ratio and affect the nutrient uptake [8]. Apart from these primary effects, plants are also exposed to oxidative stress conditions that includes the production of Reactive Oxygen Species (ROS) [5]. An increase in ROS leads to lipid peroxidation, and higher degradation of macromolecules [9]. This also causes severe damage to the cellular DNA, changes in cellular membrane permeability resulting in ion leakage from the cell [10]. Plants therefore, experience a severe decline in water potential, leading to plant water deficit in salt stress conditions. Salinity also causes inhibition of plant photosynthesis [11].

Sodium chloride toxicity is related to electrical conductivity (EC), an indicator of plant tolerance to salt stress. Most of the horticultural vegetable crops including tomato have a salinity threshold that is lesser or equal to 2.5 $dS\ m^{-1}$ [12]. Horticultural crops show adverse impacts of salinity characterized by poor seed germination and, also seedling growth retardation [13]. Inhibition of leaf growth has also been observed in plants exposed to excessive root zone salinity and has been attributed to decreased cell turgor, decreased photosynthetic activity, and activation of metabolic signals between stress perception and adaptation [14]. Salinity stress decreased the yield marketability of fruits, roots and tubers [15].

One of the most important horticultural crops cultivated on a global scale is *Lycopersicon esculentum* Mill. (Tomato). Tomato is considered as "moderately sensitive" to salinity [16]. It is a principal source of phytonutrients and bioactive compounds [17]. High NaCl concentration in soil reduces the rate of germination and delays germination of tomato seeds. Salinity decreases water uptake by root hairs from soil which hinders sugar metabolism inhibiting cell division and thus suppressing the leaf, flower and fruit [18]. Salinity is one among the most important abiotic stress factors threatening plant productivity and yield [19].

1.2 Seed priming and mechanical conditioning practices

Two strategies, seed priming and mechanical conditioning, that are considered age-old are now gaining considerable research interest in the last decade for their role in conferring tolerance to abiotic stress conditions such as salinity, drought, chilling stress. Seed priming is a simple, cost effective and practically proven technique to accelerate rapid and also, synchronized seed germination [20]. Seed priming increased yields in many horticultural crops under adverse environmental conditions [13]. Mechanical conditioning on the other hand, is a process of physical stimulation or stress deliberately applied in order to manage plant growth and quality [21]. Studies show that the mechanical conditioning method reduce plant growth by rubbing stems, brushing shoots, shaking, vibration techniques, mechanical impedance, or even by perturbing plants with either water, wind. This method also results in cross-tolerance to stress conditions [22].

2. Seed priming

Hydration of seeds triggers the germination of seeds through three phases: imbibition, activation and radicle protrusion [23]. Seed priming is a pre-sowing hydration technique applied in a controlled condition, that results in the activation of key metabolic processes during the seed imbibition phase before the radicle emergence [24]. Seed priming transforms the seeds from a metabolically inactive state into a quasi-metabolically active state. However, this quasi-metabolically state does not support the complete emergence of the radicle [25]. The pre-sowing technique is consequently followed by seed drying and maintenance of near to original weight and moisture content of the seed (10–15%) [26]. The seed priming technique has been shown to improve salinity stress tolerance in a variety of plants (**Table 1**).

2.1 Seed priming techniques

The classification of the seed priming technique depends on the chemical nature of the priming agent. The hydropriming technique is a simple, inexpensive, and eco-friendly seed priming technique that involves imbibing seeds in distilled water [41]. In the chemical priming technique, the seeds are hydrated in salt solutions such as KH_2PO_4, KNO_3, $CaCl_2$, $MgSO_4$, NaCl, KCl while, osmopriming involves the use of osmolytes such polyamines, PEG, mannitol etc. [20]. Biopriming integrates seed uptake with biologically active bacterial inoculants such as Plant Growth Promoting Rhizobacteria [PGPRs] to enhance germination and seedling vigor [36]. Hormonal priming involves the hydration of seeds in an aerated medium of various plant growth promoting hormones such as abscisic acid, kinetin, SA [salicylic acid], GA3 and ascorbate [40]. In the solid matrix priming [SMP], the matrix potential of the priming solution is controlled during seed uptake by the addition of solid matrix substances, for example compost, clay and sand that, create matrix forces to hold water and slow solute uptake by the seeds [42]. Nutripriming improves the available nutrients and water for seed preparation of the emerging plant by adding magnesium, zinc and boron, which effectively promotes germination, growth and development, early flowering, early maturity, grain filling rate and yield of multiple crops [43]. Among all the priming techniques that is reported, hydro, osmo, chemical and hormone priming are the most commonly used, though nano-priming has stimulated intense interest in the recent years [44]. In the seed priming technique, seeds are placed in a specific, defined concentration of priming agent for a specific period [45]. The success of the priming technique relies on the water potential, priming

Sl No	Priming Technique and Model Plant	Priming agent and concentration/ Duration of priming	Major beneficial effects of Priming	References
1.	Chemical Priming Model: Tomato (*Lycopersicon esculentum*)	Ascorbic acid (0 and 100 mM AsA) for 1 h	• Increases water potential and water use efficiency • Modulates Antioxidant system.	[27]
2.	Nano priming Model: Tomato (*L. esculentum*)	Carbon nanotubes and graphene (50, 250 and 500 mg/L) for 24 h	• Improves the content of bioactive components in fruit • Enhances antioxidant mechanism.	[28]
3.	Irradiation Model: Tomato (*L. esculentum*)	UVC (0.85 and 3.42kJm^{-2})	• Activation of photosynthetic processes • Accumulation of soluble proteins, phenolic compounds, flavonoids and carotenoid content	[29]
4.	Chemical priming Model: Tomato (*L. esculentum*)	KNO_3 (0.25, 0.50, 0.75, 1.0 and 1.25) for 24 h	• Improved final emergence and mean emergence time • Enhanced production of total soluble sugars and phenols	[17]
5.	Solid matrix Model: Tomato (*L. esculentum*)	4% Sand particles (diameter 0.5 mm to 2 mm) for 72 h	• Increased final germination percentage and vigor index • Enhanced the antioxidant defense system	[30]
6.	Solid matrix Model: Brocolli (*Brassica oleracea L. var. italica)* and Cauliflower *B. oleracea L. var. botrytis)*	Vermiculite and H_2O (1:1.5:2) for 2 days	• Enhanced antioxidant activities (POD, CAT) and osmolytes (Proline, soluble sugar and soluble protein).	[31]
7.	Osmopriming Model: Brocolli (*B. oleracea* L. var. *Italica)*	KCl	• Regulated glucosinolate metabolism and phenolic production	[32]
8.	Biopriming Model: Peas *Pisum sativum.* L	*Typha angustifolia* L. (leaf extract – 40 g/L) for 24 h	• Decreased MDA level and well maintanence of membrane integrity • Increased osmoprotectant production (Proline, total soluble sugars, K+ and P), Carotenoid and Chlorophyll content.	[33]
9.	Hydropriming Model: Napa cabbage *(Brassica rapa L. subsp. pekinensis)*	Distilled water for 10 h	Increased antioxidant activities such as POD CAT and osmoticum (Proline).	[34]
10.	Chemical priming Model: Tomato (*L. esculentum*)	Polyamines Putresine (25 mM) Spermine(2.5 mM) Spermidine(2.5 mM) 20 ml for 24 h	• Increased membrane integrity, photosynthetic pigments and proline • Enhanced enzymatic and nonenzymatic antioxidant responses	[35]

Sl No	Priming Technique and Model Plant	Priming agent and concentration/ Duration of priming	Major beneficial effects of Priming	References
11.	Bio-Priming Model: Okra *(Abelmoschus esculentus)*	*Enterobacter hormaechei* sp. (PGPR) for 1 h	• Improved germination parameters and seed vigor index • Increased K^+ and P^+ uptake • Increased chlorophyll index	[36]
12.	Hydro, chemo and hormonal priming Model: Mustard (*Brassica juncea* L.)	Water, CaCl2 (100 μM), ABA (100 μM) for 18 h	• Increased rate of germination • Enhanced SOD and GPX activity • Decreased oxidative damage	[37]
13.	Chemical and hydro Model: Sunflower (*Helianthus annuus* L.)	KNO3 (500 ppm) for 2 h and water for 18 h	• Enhanced germination characteristics, root and shoot length • Decreased mean germination time and abnormal germination percentage	[38]
14.	Chemical and Hormonal priming Model: Periwinkle *(Catharanthus roseus* L.)	KNO3 (1%) Salicylic acid (0.5 mM) for 15 h	• KNO_3 alleviated the growth of seedlings, mitotic activity and decreased chromosomal abnormalities • SA increased plumule dry weight and ratio of plumule weight: root	[39]
15.	Chemical, osmo and hormonal priming Model: Maize *(Zea mays)*	NaCl (50, 150 and 250 mM) PEG (10, 15 and 20%) GA (5, 10 and 15 mg/L) - for 24 h	• Recommended priming concentrations were found to be: • NaCl - 50 mM • PEG – 15% • GA – 10 mg/L • Increased shoot biomass (NaCl priming) • Increase grain weight (Water, NaCl and PEG)	[40]

Table 1.
Seed priming technique leading to improvement in salt tolerance in different plant species.

duration, priming agents, and also seed condition that influence subsequent seed germination and seedling emergence. Therefore, optimizing the priming technique and priming duration has been an active area of research investigation for improving seedling establishment and plant productivity under a variety of environmental conditions [45].

2.2 Mechanism of seed priming induced salinity stress cross-tolerance

2.2.1 Seed water uptake and seed physiology

Seed priming ensures maintenance of optimal moisture levels before sowing through controlled hydration process. It allows physiochemical processes (metabolic activities) to be carried out in treated seeds prior to germination, but limits the emergence of radicle prior to sowing. The seed priming process is accomplished in three phases [24]. The first phase is phase of imbibition where uptake of water is controlled in primed seeds [23]. Controlled hydration takes place in the first phase, limiting the seed moisture sufficient to initiate early phase of germination [23]. This is subsequently followed by the activation phase, in which a series of DNA repairing and metabolic activities commences at the cellular level. In this phase, synthesis of proteins, activation of enzyme, antioxidant system and DNA repair takes place [24]. Priming is thought to increase the activity of several enzymes involved in metabolism of carbohydrates (alpha and beta amylases), proteins (proteases) and lipid mobilization that are involved in the stored reserve mobilization. Seed priming also permits early DNA replication and repair, increased RNA and also decreases the leakage of metabolites [46]. Priming also enhances antioxidant synthesis for re-orienting the seeds to defend against oxidative damage [47]. In the third stage, the uptake of water is rapid and protrusion of the radicle commences [24]. This is then followed by the dehydration process or drying. The seed drying procedure is critical as it can affect seed vigor and longevity and needs ideal conditions for seed storage [23]. The ideal conditions for maintenance of seed integrity are usually done by bringing the seeds to its original moisture content (**Figure 1**). Varier et al. [48] reported that, since the seed priming technique leads the seeds to advanced physiological status in comparison to unprimed seeds, the primed seeds are more prone to seed deterioration, and therefore a major bottle-neck in the commercialization of primed seeds. However, recent studies show that mimosine, a chemical inhibitor of cell cycle, identified from a large screen of biologically active compounds, resulted in the development of green cotyledons and prevented seed deterioration after the seed priming technique. The identification of a cell cycle inhibitor suggests that the cell-cycle is an important checkpoint in maintenance of seed integrity during priming. However, to be applicable commercially, the optimization of concentration of chemicals for seed priming needs to be carried out [49].

Seed priming mainly improves salinity stress tolerance through cross-tolerance, through two main mechanisms. In the first well- understood mechanism, seed

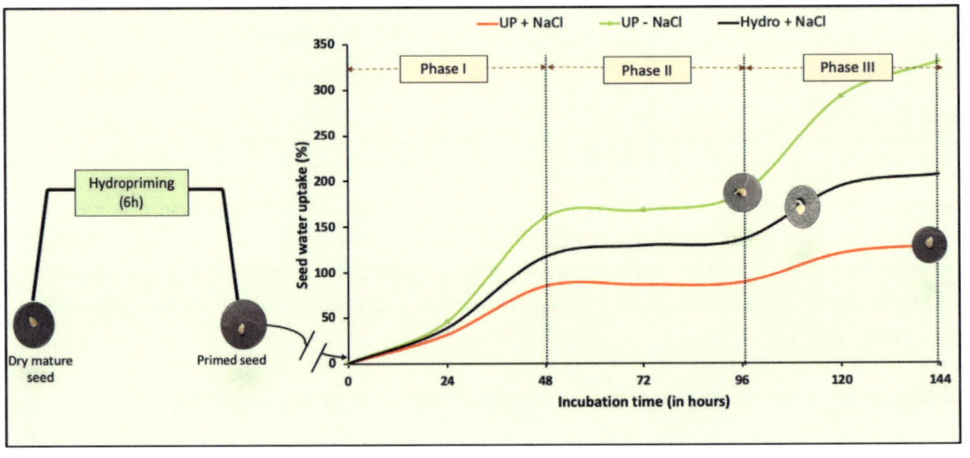

Figure 1.
Rate of seed water uptake (%) in the three phases of seed germination after exposing the hydroprimed and unprimed seeds of Lycopersicon esculentum mill. To 100 mM NaCl stress conditions. Tomato seeds (Lycopersicon esculentum mill.) belonging to variety Arka Rakshak was procured from the Indian Institute of Horticultural Sciences (IIHR, Hesaraghatta, Bengaluru, Karnataka). Dry seed lots were weighed initially. Healthy seeds were surface sterilized using routine methods. Sterilized seeds were imbibed in distilled water for 6 h and subsequently dried (room temperature). Seeds were then allowed to germinate in petri plates layered with filter paper soaked in distilled water or 100 mM NaCl alone. Each treatment (unprimed or hydroprimed) consisted of 60 seeds.

priming activates germination through enhanced energy metabolism, early reserve mobilization, embryo expansion and endosperm weakening. Secondly, seed also priming leads to abiotic stress that represses the radicle protrusion. This re-programming results in "priming memory" in seeds. The "priming memory" is mainly epigenetic in nature. These two mechanisms together mediate the improved stress tolerance of the primed seeds [25].

2.2.2 Seed priming induced salt stress response in plants

Salt stress affects plants leading to both osmotic and ionic stress. High concentration of Na^+ and Cl^- ions in the environment causes plant water deficit. The osmo protectants such as proline, amino acids, soluble sugars and proteins can regulate the cellular osmoticum in response to the reduced external water potential [23]. Seed priming is shown to improve osmotic adjustment when plants are exposed to salt stress condition. Yan 2016 conducted hydro priming of *Brassica rapa* subsp. *pekinensis* cultivar Xiaoza 56′ seeds which is considered salt sensitive (Napa cabbage). The seeds were hydroprimed for 10 h and germinated under different NaCl concentration (0, 50, 100, 150, 200 or 250 mM). Hydropriming increased the germination percentage and early seedling growth which was however, delayed in unprimed seeds grown under similar NaCl conditions. There was a significant increase in proline in 100, 150 and 200 mM NaCl concentrations, resulting in reduction in osmotic stress. However, in 50 mM NaCl there was a decrease in proline content. This study also concluded that an increase in salinity also gradually increased the proline contents [34]. Similar results were also obtained by biopriming Peas (*Pisum sativum*). Biopriming of *Pisum sativum* L, var. Lincoln with extract of *Typha angustifolia* for 48 h and exposed to 120 mM NaCl concentration resulted in enhancement in osmoprotectants such as proline, total soluble sugars along with K^+ ion concentration [33]. Therefore, seed priming regulates the osmotic potential and alleviates salt stress conditions by improving water absorption by the cell, aiding seed germination [50].

Accumulation of Na^+ ions under saline condition results in ionic stress, causing the structural damage to cell membrane and inhibiting embryo activity [51]. Increase of Na^+ in the cytoplasm also hinders the entry of other ions particularly K^+ which affects the physiology in plants [51]. Inorganic ions such as Ca^+ mitigates the negative effect of Na^+ and maintains ion homeostasis under NaCl stress conditions. This was corroborated from research performed in *Sorghum bicolor* (L.) Moench [51]. In this study, $CaCl_2$ was used as priming agent to ameliorate salt stress toxicity in sorghum. It was shown that $CaCl_2$ priming resulted in upregulation of antiporter genes such as NHX_2, NHX_4, SOS1 and the K^+ transporters AKT_1, AKT_2, HKT_1, HAK_1 and KUP which may be responsible for decreasing the Na^+ levels and increasing of K^+ levels in tissues [51]. Calcium is also a vital element for maintenance of cell-wall structure, cell elongation and cell-division [23]. Ben Youssef [52] reported that NaCl negatively affects the Membrane Stability Index (MSI) (37%) and disrupts the membrane integrity. MSI is a physiological index that estimates the percentage of membrane injury based on the electrolyte leakage [53]. Further, the authors have shown that priming with $CaCl_2$ **decreased MSI in** *Hordeum vulgare* **(L. Manel) and** *Hordeum maritimum* **seedlings** [52]. Calcium chloride signaling also plays a vital role in activating SOS pathway by directing Na^+ efflux [51] resulting in minimal leakage of ions from the plant cell and also, decreases the extent of damage to the cell membrane.

Salinity also induces membrane damage by the accumulation of inhibitory levels of reactive oxygen species (ROS) [54]. ROS scavenging can be achieved by two complex systems: the enzymatic comprised of superoxide dismutase (SOD), peroxidase (POD), catalase (CAT) and the nonenzymatic consisting of ascorbic acid, alkaloids, carotenoids and tocopherols [54]. In a case study, solid matrix seed priming (using vermiculate and water) of *Brassica oleracea* var. *italica* (Broccolli) increased the SOD, POD, CAT activities along with proline, sugars and proteins [31]. These studies demonstrate that ROS homeostasis is critical for plant responses to salt stress conditions. There is a negative relationship between antioxidant activity and MDA levels. Seed priming reduced MDA by enhancing antioxidant activity [25, 30].

2.3 Amelioration of salinity by seed priming

Seed priming is as cost-effective practice in establishing uniformity of germination [55]. In an experiment conducted in our laboratory with tomato, the dry seeds were primed in distilled water for 6 h (hydropriming). The dried hydroprimed seeds were allowed to germinate in control, unstressed conditions (0 mM NaCl). Seeds undergo re-imbibition and enters into phase I of the seed water uptake process during which rapid initial water uptake takes place, triggering all cellular processes. Imbibition of water takes place rapidly in control tomato seeds as they are not exposed to NaCl stress conditions. However, there was a decline in the percentage of water absorption in the seeds exposed to NaCl stress (100 mM) (84.5%) in comparison to control seeds (159.9%) at 48th h. In tomato, the first phase was shown to last from 0th h- 48th h, after imbibition. However, hydroprimed seeds that were exposed to NaCl conditions (100 mM) showed seed water content of 116.9% (at the 48th h) water in comparison to stressed seeds during phase I (0–48 hr). This confirms that hydropriming of tomato seeds improves the water uptake mechanism in tomato (**Figure 1**). The second phase, the lag phase, of the seed germination process is characterized by a plateau where there is limited water uptake, but increased metabolic and cellular activity. This phase is marked by metabolic activities such as protein synthesis, activation of antioxidant system and enzymes, DNA repair mechanism and ATP production [24]. All macromolecules such as carbohydrates, proteins and lipids are mobilized by enzyme to nourish the newly developing embryo. For instance, α- amylase activity and total soluble sugars increased in wheat, when primed with benzyl amino purine and sorghum water extract [56]. Further, with the onset of favorable environmental conditions in nature, these nutrients are redirected to the next phase (Phase-III) of seed germination i.e., radicle emergence. Hydroprimed seeds exposed to NaCl stress (100 mM) exhibits 135.3% of seed water uptake when compared to unprimed (88.1%) seeds grown under similar conditions (100 mM NaCl). The surge in water uptake resulted in hydrolysis of macromolecules aiding in early radicle emergence in comparison to unprimed stressed seeds (**Figure 1**). It is now understood that with seed priming, the lag time of imbibition is reduced [57]. The swelling of the embryo in primed seeds speeds up the germination by facilitating water absorption [55]. Seed priming also reduced physical resistance of the endosperm, stimulates pre-germinative re-orientation of metabolic processed and leaches out the chemical inhibitors of germination [58]. In summary, the seedlings emerge faster, show vigor and performs better in NaCl stress conditions (**Figure 1**). In another experiment, seeds of *Lycopersicon esculentum* (tomato) of two varieties (var.205 and var.206) were primed with sand (solid matrix priming) (4%) at 25°C for 72 h. The treatment improved germination by increasing the antioxidant enzyme activities in laboratory conditions. *Lycopersicon esculentum* var.206 showed higher activities of CAT, POD,

SOD, APX (4.5, 15.3, 6.6 and 4.1% respectively) with high salt stress (150 mM) in comparison with var.205. The increase in the activities of antioxidant enzymes is understood to improve the innate defense mechanism in the primed seeds leading to enhancement in the seed vigor in tomato [30].

Garcia et al. reported that nanoparticles can be used as effective priming agents in tomato. The authors showed that nanopriming of tomato seeds with Graphene based-Graphene 500 (GP500) and carbon based-Carbon nano tube 500 (CNT 500) for 24 h resulted in enhancement in chlorophyll content by 8.75 and 13.3% when compared with untreated seeds in 50 mM NaCl stress conditions. The authors reported that the ion leakage decreased by 10.71% with the addition of CNT 500 when compared to non-sonicated control. All these studies suggest that priming with nanoparticles can cause subtle to marked changes in biochemical, physiological, and biological properties. The carbon nanomaterials (priming agent) penetrate the seed coat by forming new pores allowing the entry of water, oxygen, nutrients into the cytosol. This facilitates the seed germination process [28].

Further, green house experiment conducted in tomato with two different concentrations of ascorbic acid (hormonal priming; AsA-0 and 100 mM) for an hour enhanced leaf water potential (LWP) and water use efficiency (WUE) in NaCl stress conditions. LWP increased by 25% when seeds were primed with 100 mM AsA exposed to NaCl (100 Mm) and showed lower LWP in the absence of NaCl. Water Use Efficiency (WUE) increased by 37.7 and 21.6% with 0 mM and 100 mM NaCl. There was a significant interaction effect of NaCl and AsA within and between NaCl concentrations. LWP is directly correlated to WUE as water availability affects stomatal opening. The dry mass of leaves and fruits increased with AsA priming by 19.5 and 17.9% respectively in plants subjected to NaCl [27].

Abdel-Aziz et al. 2019 demonstrated the positive effect of nanopriming on the growth and productivity of *Phaseolus vulgaris* cv. Contender (French bean). The authors investigated the possible effects of two different methods (foliar spray and seed priming) using nanoparticles on French bean. The two nanomaterials nanochitosan (Cs) and carbon nanotubes (CNTs) were applied either by foliar spray or by seed priming technique. There was significant increase in the biochemical composition of yielded seeds of foliar spray than in seed priming in comparison to control seeds. The days to harvest decreased to 80 days through foliar spray without compromising the yield, whereas seed priming resulted in 110 days to harvest. Therefore, it was concluded that Cs nanoparticles improved growth and yield more than CNTs through foliar application. The differences in stress responsive effects in *Phaseolus vulgaris* between the two modes of application (foliar vs. seed priming) was explained on the basis that, nanoparticles are best absorbed by the stomata when they are foliar sprayed, providing more nutrients [59].

Alamer et al., evaluated the efficiency of seed priming by irradiating tomato seeds with a dose of 0.85kJm^{-2} and 3.42kJm^{-2} under laboratory conditions. UV-C priming enhanced root and leaf biomass by 51 and 36% with 0.85 kJm^{-2} and 10 and 25% with 3.42kJm^{-2} respectively. UV-C priming with 0.85 kJ m^{-2} and 3.42kJm^{-2} also decreased Na^+ content by 33 and 20% in roots and 30 and 40% in leaves respectively. K^+ accumulation on the other hand, decreased in roots by 38% (with 0.85kJm^{-2}) and 22% (with 3.42kJm^{-2}) compared to untreated controls. Interestingly, treatment with 0.85kJm^{-2} alleviated NaCl-induced stress by maintaining root biomass, root potassium supply and leaf protein content in comparison to controls. On the contrary, treatment with 3.42kJm^{-2} stimulated the NaCl resistance of the tomatoes by maintaining the protein, total polyphenol and tannin content of the roots in comparison to controls. This, points to the fact that different concentration of priming agents can influence the resistance of plants to salinity stress to varying degrees [29].

3. Mechanical conditioning methods for stem length reduction

The principal benefit of mechanical conditioning strategy for farm application is to produce plant transplants that are strong, more elastic, sturdier and shorter. The plant stem supports the weight of branches, fruits and therefore, requires adequate strength to withstand wind, lodging, rain and other environmental stress damages. Chemicals that are widely used as plant growth inhibitors have the drawback of toxicity and also, continuous inhibition [60]. Mechanical conditioning in plants is therefore, a much effective method to control and regulate the plant height. Among all the methods used for mechanical conditioning, brushing treatment has received the most significant research interest in vegetables crops due to the ease of its application and the possibility of automation for field applications. Brushing provides for the tactile stimulation of the plant growing points. In most of the research reported, brushing provides a mechanism to reduce plant height, but is has also been shown to increase stem and petiole strength. The device used for brushing treatments must also be robust and strong enough to manipulate the shoots in a plant. In an interesting study, experiments were performed to compare chemical treatment using paclobutrazol (a chemical bases plant growth retardant) with physical stimulation (40 brush strokes per day) or cold air (air flow of 2 m sec^{-1}, at 18°C each day). Both paclobutrazol and mechanical stimulation provided the best results in terms of stem reduction of 25% in comparison to control plants. Brushing treatment also resulted in reduced plant biomass. However, paclobutrazol treated seedlings grew better than the brushed or cool air flow treated seedlings after they were transplanted to the field [61].

Mechanical stimulus (MS) has also been done by gentle rubbing. For instance, mechanical stimulus was applied to 3-week tomato plants at the 4th internode by mechanical rubbing between the thumb and fore-finger and rubbed back and forth, once for 10s. Internode length was measured thereafter after 14 days. Results showed that rubbing of internodes resulted in significant reduction in the elongation of the stressed internode (4th) and also, the neighboring internode (5th internode). This was also corroborated with increased lignification (7.68% in the control to 10.08% in the mechanically stressed sample). It was reasoned that the lignin accumulation observed with mechanical stimulation may function as a plant growth inhibitor. The authors also showed that the response of tomato plant to mechanical stress by the inhibition of internode elongation was related to the induction of CAD activity and peroxidase isoforms [62]. Further, a reduction in indole-3-acetic acid (IAA) content was also detected in the rubbed internode and the upper internode. These results also suggested that a decrease in rubbed internode length is a consequence of IAA oxidation leading to increase in enzyme activities (PAL, CAD and POD) [62]. Further, it was also shown that MS treatment through rubbing increased stem resistance to tensile forces. The mean tensile strength resulted in a significant increase of 113% in rubbed plants, when compared to plants that were not treated [63]. It was also shown that calcium induced changes brought about by rubbing was thought to elicit downstream changes in gene expression and that, the Ca^{++} effect is modulated by calmodulin [64].

Apart from these methods, mechanical impedance is a well applied method in plant mechanical conditioning to control excessive stem elongation. In a study conducted, tomato plants grown in green house was used to compare between mechanical impedance and brushing treatments. Mechanical impedance procedure was applied by means of a 5-mm thick acrylic sheet (Plexiglass) with a mean pressure of 66 N/m^2. On the other hand, brushing experiments were done using a Styrofoam by stroking 20 times. Results from these studies showed that physical impedance method was equally suited for stem reduction in tomato plants just, as the brushing method. In this study, the authors noted that both the treatments (impedance as well as brushing) reduced stem length by 3–4 cm. This reduction in stem length corresponded to a 40% reduction rate in elongation during 7–10 days of plant treatment. Impedance resulted in shorter and thicker stems, and also caused more horizontally oriented leaves [65]. It was also shown from these green-house experiments that, physical application by mechanical impedance below 66 N/m^2 was not effective in decreasing plant height. Physical impedance was also shown to result in increased stem diameter and adventitious root formation at the base of the stem. But, this method of application is more laborious than brushing experiments and requires more equipment to regulate stem height control [65]. There are also other studies that compared the two treatments (Brushing and Mechanical impedance). In another study, mechanical conditioning treatments were begun when the seedlings were 6 cm tall and 17 d old. The brushing treatment was applied with an unpainted, 25-mm-diameter hardwood dowel pulled gently 20 times, back and forth, across the canopy each day for 15 d. The impedance treatment was applied by suspending an acrylic sheet (4 mm thick) just below canopy height overnight. Mechanically conditioned transplants of processing tomatoes resumed growth after transplant shock as quickly as did untreated plants, and subsequent canopy development was also equal. In 4 years of field trials, yield was not reduced by mechanical conditioning. In this study, neither earliness nor defects in the fruits of the first cluster were affected by mechanical conditioning. Early and total yields were equal in both years that fresh-market crops were tested. Thus, there were no adverse effects on the field performance of either processed or fresh-market tomatoes as a result of reducing stem elongation by mechanical conditioning before transplanting [65].

In other studies, wind treatments were performed using a 20-inch fan blade, that were suspended 28 inches above tomato transplant seedlings. Wind speed was provided at the average speed of 30 km/hr. From this study, it was shown that tomato seedlings showed a 19% reduction in stem elongation with wind treatment in comparison to control seedlings. Interestingly, all day or all night or all day/night reduced the internode length compared to control tomato seedlings. In contrast, other times of the day/night cycle, for instance (early, mid or late day/night) wind treatments did not cause any alteration in internode length when compared to control tomato seedlings. The authors concluded that the impact of wind treatments vary greatly diurnally [66]. Further, Sparke et al. demonstrated the usefulness of an automated, directed air-stream application to control plant height in tomato and other ornamental plants. The air stream applied to the plants was generated by a stationary compressor with downstream connected pressure regulator. The automated air-stream application on a metal frame suspended on a longitudinal guide rail and was equipped with an electric motor to facilitate forward and backward movements across the entire greenhouse. The authors observed that directed air-jets applied either as laminar air stream or turbulent free air streams resulted in 26–36% reduction in tomato plant height [67].

A similar result was observed in the height of tomato seedlings by wind-blowing [68]. The authors also observed that the wind treatment of 0.6 m/s every 30 minute for 5-minute reduced tomato seedlings hardness and elastic modulus by about 25 and 24%, respectively. Similar improvement in plant growth characteristics have been shown using Bending/Flexure of plant and also, vibrations [1].

3.1 Mechanical conditioning and other plant responses

3.1.1 Seed germination

Mechanical stimulation has also been used as a simple, cost-effective technique to break seed dormancy and improve seed germination. The most popular techniques to improve seed germination are sound and ultrasonic waves and vibration. Vibration is described by two parameters of application, the frequency and amplitude. In *Arabidopsis thaliana*, the effect of sinusoidal vibration (40–120 Hz) on seed germination and amplitude equal to or smaller than 0.42 mm was studied. The authors observed that amplitude of 0.42 mm and frequencies higher than 70 Hz or amplitudes larger than 0.33 mm and frequencies of 100 Hz increased *Arabidopsis* seed germination rate. By employing ethylene in sensitive mutants, it was concluded by the authors that, mechanical stimulation by vibration increased the rate of seed germination most likely through the action of ethylene [69]. In another study, Yang et al. showed an increase in seed germination rate and potential by 15 and 14% respectively in tomato using 40 KHz ultrasound frequency [70].

3.1.2 Mechanical conditioning and cross tolerance to salinity

Plants exposed to various forms of mechanical stress conditions caused by wind, rain, herbivores, and pathogens induce responses in the plant that were shown to have an adaptive value. In earlier studies, mechanical stress associated with damage or wounding was shown to result in increased resistance to insects, pathogens. For instance, Capiati et al. showed that mechanical wounding performed by cutting with a dented forceps improves salt stress tolerance in tomato plants. Wounded tomato plants were shown to be more tolerant to salt stress conditions than un-wounded plants. Pre-wounding also resulted in increased relative water content in comparison to un-wounded leaves. The results further point to the role of Calmodulin like activity (LeCDPK1), a Ca^{2++}-dependent protein kinase from tomato in salinity cross-tolerance. It was concluded that pre-wounding of tomato increased salt stress tolerance through the mechanism involving the signaling peptide systemin and the subsequent synthesis of Jasmonic Acid, leading to increased expression of LeCDPK1 [71]. Further, it was also shown that artificial wounding experiments induce defense response in tomato by involving the crucial involvement of a signaling molecule, H_2O_2. A sudden surge in H_2O_2 levels was shown within the few hours of wounding stress performed on either the leaf mid-rib or lamina of tomato. Primed, wounded plants showed increased total phenol, total flavonoid content and antioxidative activity. Alleviation of salt stress was higher in mid rib cuts than in lamina was confirmed through the stabilization of relative water content and also, an increase in antioxidant scavenging activity. Therefore, the defense response was stronger in plants with plants with leaf midrib injury than in those with laminar injury [63]. However, these experiments were done by methods that lead to discernible injury in the plant. The question remains if the salinity stress cross tolerance can be initiated

SI No	Primary Stress	Plant	Cross Adaptation	Conditions	Reference
1.	Wounding	Tomato	Salt tolerance	Controlled	[71]
2.	Brushing	Tomato	Cold tolerance	Controlled	[73]
3.	Weight Loading/Pressing	Wheat	Cold tolerance	Semi-field	[74]
4.	Cylinder Roller	Wheat	Cold tolerance	Controlled	[75]
5.	Heat Shock	Maize	Heat, Chilling, Drought and Salt	Controlled	[76]
6.	Sound vibration	*Arabidopsis*	Drought	Controlled	[77]
7.	Sound vibration	*Mentha pulegium*	Salt stress	Controlled	[78]
8.	Rubbing/stroking	Bean	Drought	Controlled	[79]
9.	Salt	Tomato	Wound stress	Controlled	[80]
10.	Increasing rotational speed	Tobacco	Heat, chilling, salt	Controlled	[22]

Table 2.
Type of mechanical stress induced cross adaptation in plants.

by conditioning methods without causing explicit damage to the plant tissues. Results from other studies show that gentle sweeping of leaf surfaces leads also to strong resistance to *Botrytis cinerea*. The soft mechanical stress induced burst of calcium flux and ROS induce the expression of genes required for defense against the virulent fungus *B. cinerea* [72] (**Table 2**).

3.1.3 Improvement in metabolites and nutritional quality

In a study conducted at an experimental station using floating hydroponics, vegetable Horti crops (lettuce and chicory) were given mechanical conditioning (MC) treatments. This was performed by brushing the vegetable crops with a burlap cloth either for 10 passes (MC10) or 20 (MC20) per day [81]. In this study, all the specialized metabolites (Ascorbate, Total Phenolic and Total flavonoid content) showed a marked increase in amount in MC20 treatments in comparison to control. In general, the highest mechanical conditioning treatments (20 passes) resulted in antioxidant enhancement in both lettuce and chicory. Mechanical conditioning treatments in this study therefore, resulted in major alterations in metabolic pathways (phenylpropanoid) leading to increase in specific activation of phenolic compounds (TPC and TFC), pigments (chlorophyll and carotenoids) and also antioxidants (Ascorbate), those, that are significant for defense response. It is well established that exposure to brushing or wounding in nature by herbivores, insects etc. cause major changes in the polyphenolic content [22]. The increase in the content is crucial for plants to recover from any damage to DNA and cell membrane from the rapid accumulation of ROS in plant cells. Antioxidants function as radical scavengers, chelators, quenchers and oxygen scavengers [82]. Most interestingly, the total chlorophyll content and carotenoids also increased with exposure to MIS in lettuce. This is a significant observation and therefore, it can be concluded that imposition of MS as a good practice in horticultural crops for production of plants with high nutritional rich phytochemical content [81]. There is a large body of scientific evidence that consumption of nutritionally enhanced food crops/vegetables could have a beneficial effect on human health [83].

4. Mechanical conditioning and mechanism of action

Plants are sessile and, in order to respond to mechanical stimuli in the environment, plants begin an intricate cascade of signaling events. The signaling cascade is stimulated with a receptor, that perceives the mechanical stimulus, resulting in a biochemical reaction, involving Ca^{++} and culminates in the thigmotropic response. Plants do not possess specialized cells for thigmotropic perception. Each of the individual cells is understood to have the ability to sense MS that gives rise to a cellular response. In *Arabidopsis thaliana*, trichomes and passage cells can sense mechanical stimuli [84]. In an elegant experiment carried out in *Arabidopsis thaliana*, it was demonstrated that plants sense such mechanical stimuli through changes in cytosolic Ca^{++}. The authors showed that plants can sense pressure changes applied onto the leaves using micro-cantilever device that exert a compressive force [84]. The plant thigmotropic response is also shown to be dose dependent, saturable and also systemic, in that the response is shown to translocated from the mechanostimulated local tissue to unperturbed distal tissues [85]. This is now understood to be mediated by proteinaceous as well as diffusible signaling molecules [86].

4.1 Calcium and TCH genes

Calcium acts a universal signaling molecule essential for the response to adaptation to a changing environmental condition. The cell perceiving a stress stimulus leads to a rapid rise in cytosolic Ca^{2++} through the opening of calcium channels present in the plasma membrane and intracellular organellar membranes. This quick and transient rise in cytosolic calcium is a key factor in expression of stress-responsive genes and physiological responses of plant cells to stress conditions. In part, the morpho-physiological response to mechanical stimuli is due to alterations in gene expression and synthesis of new proteins. This process of mechanoreception was elucidated by the identification of TCH genes (Touch) genes, induced within 10–30 minutes of exposure to perturbation by touch or wind [87]. Several genes have been identified TCH1, TCH2 and TCH3 that encode calmodulins or calmodulins like proteins. The TCH4 gene encodes a xyloglucan endotransglucosylase (XTH) involved in cell wall modification [88]. The discovery of calmodulins or calmodulins like proteins as universal calcium-dependent activators of enzymes and their activation with Mechanical stimulation suggests the involvement of Ca^{2++} in mechanosensing. Interestingly, transcriptomic studies have shown that mechanostimulation results in up-regulation of large number of *Arabidopsis* genes (including genes for calcium binding, disease resistance and cell-wall modifying proteins) [89]. This also suggest a central role of gene regulatory network in the thigmomorphogenetic response.

4.2 Mechanosensing

Mechanosensing in plants is understood to involve a large class of mechanoreceptors, among which are the mechanosensitive ion-channels, a protein complex (MS channels). These ion channels span the membrane facilitating the regulated movement of ions upon mechanical stimulation. Among these, the mechanosensitive channel of small conductance like-MSL are understood to be involved in mechanosensitive properties in *Arabidopsis* [90, 91]. *Arabidopsis* has a minimum of 10 MSL proteins, differentially localized, some in the plasma membrane, plastid or mitochondria. Another major class of proteins involved in mechanosensing and mechanoperception are the Plasma membrane localized receptor like kinases (RLKs). Plant Receptor like kinases (RLKs) recognize the damage associated molecular patterns (DAMPs) that are produced by plants during pathogen attacks and induce a series of intracellular events such as MAPK cascade, ROS burst and Ca^{2++} spiking [92]. Ca^{2++} spiking immediately after mechanical stimulation is one of the upstream signaling events in the plant that triggers rapid protein phosphorylation through calcium-dependent protein kinases (CDPKs) and calcium binding proteins (CBPs). CDPKs and CBPs play important roles in converting Ca^{2++} signals into transcriptional responses [93].

5. Automation of MS treatment for potential benefits

One of the bottlenecks for field applications is the tangible nature of Mechanical Stimulation stress treatments. MIS *per se* is laborious and the manual nature of this technique due to deliberate human interventions can lead to varied results in the field scenario. Therefore, few attempts have been made to automate the method for field applications. One of the earliest applications was the use of low cost, manually operated brushing apparatus. This moveable system is mobile, set on castors and has a size of dimensions of about 6 × 5 feet with adjustable working height and width. Brushing of tomato seedlings were performed by pulling a four-bar apparatus across the tomato seedlings several times [94]. In yet another interesting case-study, various herbs (Basil, coriander, mint) were mechanically stimulated in production house style green-house with a machine using light clothes attached to an irrigation boom continuously for 108 times per day. Results from these experiments showed that the brushed plants significant reduction in elongation and more stability [95]. Several other production style methods have developed that include soft clothes [96], motion vibrators [97], roll-table system [98]. More recently, in a well-devised experimental design, the authors used two different robotic platforms for investigating the potential of automation in a MS study [99, 100]. In the first case, a seven-DoF Franka Emika robotic arm, equipped with a stroking end-effector consisting of a row of plastic strings, was placed in a greenhouse setting. In the second case, an autonomous cultivation bed (LOMAS^{++}) with a growing area of 1.2 m^2 and, capable of holding plants of 30–35 com was also equipped with fully automated three-DoF robotic manipulator with a stroking end effector. In the experiment 1, the objective was to study if the frequency of mechanical motions affects plant morphology and elemental composition in plants in a greenhouse setting using the method elaborated in Case 1. A total of 50 pots with 3–5 basil seedlings in each was assigned randomly to 6 groups designed to provide variations in motion type (stroking and dipping) or motion frequency (either 100 times a day or 2 times a day). Treatments were provided to the treatment plants for a total of 27 days. Plants assigned to controls were not provided these treatments. In the experiment 2, a total of 200 plants were randomly split to five groups (treatments vs. controls) that differed in the amount of time of total application for mechanical treatments. In the experiment 3, variation in experimental design were provided by using softer material for mechanical treatment and also sampling strategy. A clear difference was shown in plant height with mechanical treatments using robotic platforms in experiment 1 and 3. Though there were no differences between control group and treatments in which stroking was applied 2 times a day, plants that were stroked 100 times a day produced 31% shorter inter-node length and 50% shorter stem length. Dipping experiments also showed a significant shortening of stem length. Experiment 2 which studied the effect of time of total application of MS showed that with increase in time of application (4 weeks), MS treatment produced shorter stems. On the other hand, results from experiment 3 revealed that MS treatments affects only internode length that developed after treatments but not already developed internodes. Therefore, the authors concluded that difference in treatment frequency, total time of MS application and stage of development affect the plant overall response to MIS treatment. Most interestingly, results with elemental composition showed a statistical increase in Magnesium in treatments when compared to control groups. Further, MS treatments resulted in increased DHA (Dehydroascorbic acid) and GABA levels. The authors concluded that increases in these metabolites is linked to build up of ROS and changes in nitrogen metabolism and aminoacid accumulation with mechanical induced abiotic stress [99, 100]. In another interesting application, a model automated touch machine was designed for the purpose of studying the effect of touch response in *Arabidopsis thaliana* [101]. The model automated touch machine was equipped with a H -shaped metal rack, a robotic metal arm equipped with hair brushes and a controller. The design automation in this study resulted in significant difference in the plant phenotype when compared to control treatments. Further, automation was used in this case to offer labor-saving and uniform touch response in comparison to human interventions. They are also useful to study the mechanism of touch response. For instance, using this experimental set-up, two proteins, MKK1 and MKK2 were shown to be crucial for bolting delay in *Arabidopsis* but not for rosette shape and area [101].

6. Farming practice, sustainable agriculture and conclusions

The most critical direction that practices of farming need to take is to incorporate strategies that combine agricultural sustainability by moderating the use of harmful chemicals and also, reducing the negative environmental effects of the farming practice. The strategies summarized in this chapter, both seed priming and mechanical conditioning of plants are cost-effective and are also easy to conduct. We can also conclude that under controlled conditions, both seed priming and mechanical conditioning have resulted in improvement in plant growth characteristics and stress cross – tolerance. Nonetheless, there are challenges that need to be overcome for wide-spread use under field conditions. The storage and short shelf life of the primed seeds are a limitation of the priming technique. The methods developed involve rapid re-drying of seeds for storage purposes at the end of treatment. However, seed drying can alter the beneficial effects of the priming agent, which are lost during storage [102]. The optimal seed priming treatment can vary depending on the plant species, variety and even, seed quality. This variability represents a major limitation of the priming method, as trials are required to determine the most appropriate strategy for each situation [103]. Mechanical conditioning applied on plants has similar disadvantages. The optimal results are dependent on species, growth stage, season and also, time-interval between treatments. Apart from these, mechanical conditioning treatments is also dependent on the intensity and frequency of treatments. Additionally, both these strategies need to be explored at the molecular level for better understanding of the complex mechanism. Nevertheless, both seed priming and mechanical conditioning can be considered as valuable strategy to improve plant establishment under adverse agroclimatic conditions without compromising on plant yield and under field conditions. An example of field-level utilization of one such strategy that is, employed in Japan is the practice of treading seedling using feet or even tractor equipped with a treading roller, leading to mechanical stimulation of Barley and Wheat, termed Mugifumi. Mugifumi is shown to increase tillers, and higher yields when compared to untreated plants. This is usually performed several times at the three-seedling stage and before internode start to appear. Results from experiments of treading performed in wheat seedlings showed 54% increase in grain weight per plant. The number of spikes and weight of whole plant also increased by 18 and 41%, respectively [104].

Acknowledgements

The authors acknowledge JAIN (Deemed-to-be University) for the continuous infrastructural support.

Author contributions

JMN wrote the draft manuscript with inputs from VV and KN. VV conducted the experiment on hydropriming in tomato. TR contributed to the initial experiments on MS in tomato. All authors have agreed to the final draft. The authors have no conflicting or competing interests.

Author details

Jithesh Mundaya Narayanan*, Vishwini Viswanathan, Tulasi Tirumalai Ramanujam and Kruthika Nagendra Rao
School of Sciences, Department of Biotechnology, Jain (Deemed-to-be University), Bengaluru, Karnataka, India

*Address all correspondence to: mn.jithesh@jainuniversity.ac.in

References

[1] Kouhen M, Dimitrova A, Scippa GS, Trupiano D. The course of mechanical stress: Types, perception, and plant response. Biology. 2023;**12**:217. DOI: 10.3390/biology12020217

[2] Hussain T, Koyro H-W, Zhang W, Liu X, Gul B, Liu X. Low salinity improves photosynthetic performance in Panicum antidotale under drought stress. Frontiers in Plant Science. 2020;**11**:481. DOI: 10.3389/fpls.2020.00481

[3] Jacobo-Velázquez DA, González-Agüero M, Cisneros-Zevallos L. Cross-talk between signaling pathways: The link between plant secondary metabolite production and wounding stress response. Scientific Reports. 2015;**5**(1):1-10

[4] Roșca M, Mihalache G, Stoleru V. Tomato responses to salinity stress: From morphological traits to genetic changes. Frontiers in Plant Science. 2023;**14**:1118383. DOI: 10.3389/fpls.2023.1118383

[5] Wani SH, Tripathi P, Zaid A, Challa GS, Kumar A, Kumar V, et al. Transcriptional regulation of osmotic stress tolerance in wheat (Triticum aestivum L.). Plant Molecular Biology. 2018;**97**:469-487

[6] Zhang A, Liu Y, Wang F, Li T, Chen Z, Kong D, et al. Enhanced rice salinity tolerance via CRISPR/Cas9-targeted mutagenesis of the OsRR22 gene. Molecular Breeding. 2019;**39**:1-10

[7] Kotula L, Garcia Caparros P, Zörb C, Colmer TD, Flowers TJ. Improving crop salt tolerance using transgenic ap-proaches: An update and physiological analysis. Plant, Cell & Environment. 2020;**43**:2932-2956. DOI: 10.1111/pce.13865

[8] Zulfiqar F, Nafees M, Chen J, Darras A, Ferrante A, Hancock JT, et al. Chemical priming enhances plant tolerance to salt stress. Frontiers in Plant Science. 2022;**13**:946922. DOI: 10.3389/fpls.2022.946922

[9] Gill SS, Tuteja N. Reactive oxygen species and antioxidant machinery in abiotic stress tolerance in crop plants. Plant Physiology and Biochemistry. 2010;**48**(12):909-930

[10] Godoy F, Olivos-Hernández K, Stange C, Handford M. Abiotic stress in crop species: Improving tolerance by applying plant metabolites. Plants. 2021;**10**:186

[11] Ibrahim EA. Seed priming to alleviate salinity stress in germinating seeds. Journal of Plant Physiology. 2016;**192**:38-46. DOI: 10.1016/j.jplph.2015.12.011

[12] Machado RMA, Serralheiro RP. Soil salinity: Effect on vegetable crop growth. Management practices to prevent and mitigate soil salinization. Horticulturae. 2017;**3**(2):30

[13] Zulfiqar F. Effect of seed priming on horticultural crops. Scientia Horticulturae. 2021;**286**:110197

[14] Kiferle C, Gonzali S, Beltrami S, Martinelli M, Hora K, Holwerda HT, et al. Improvement in fruit yield and tolerance to salinity of tomato plants fertigated with micronutrient amounts of iodine. Scientific Reports. 2022;**12**(1):14655

[15] Behera TK, Krishna R, Ansari WA, Aamir M, Kumar P, Kashyap SP, et al. Approaches involved in the vegetable crops salt stress tolerance improvement: Present status and way ahead. Frontiers

in Plant Science. 2022;**12**:787292. DOI: 10.3389/fpls.2021.787292

[16] Singh J, Sastry ED, Singh V. Effect of salinity on tomato (Lycopersicon esculentum mill.) during seed germination stage. Physiology and Molecular Biology of Plants. 2012;**18**:45-50. DOI: 10.1007/s12298-011-0097-

[17] Ali MY, Sina AAI, Khandker SS, Neesa L, Tanvir EM, Kabir A, et al. Nutritional composition and bioactive compounds in tomatoes and their impact on human health and disease: A review. Food. 2020;**10**:45

[18] Guo M, Wang X-S, Guo H-D, Bai S-Y, Khan A, Wang X-M, et al. Tomato salt tolerance mechanisms and their potential applications for fighting salinity: A review. Frontiers in Plant Science. 2022;**13**:949541. DOI: 10.3389/fpls.2022.949541

[19] Zaman M, Exner M. Combatting soil salinisation using nuclear techniques: The IAEA commemorates 2021 world soil day. In: World Soil Day. International Atomic Energy agency. Glasgow, Scotland: IAEA; 2021

[20] Jisha KC, Vijayakumari K, Puthur JT. Seed priming for abiotic stress tolerance: An overview. Acta Physiologiae Plantarum. 2013;**35**:1381-1396

[21] Latimer JG. Mechanical conditioning for control of growth and quality of vegetable transplants. Hort Science. 1991;**26**(12):1456-1461

[22] Li ZG, Gong M. Mechanical stimulation-induced cross-adaptation in plants: An overview. Journal of Plant Biology. 2011;**54**:358-364

[23] Khan MO, Irfan M, Muhammad A, Ullah I, Nawaz S, Khalil MK, et al. A practical and economical strategy to mitigate salinity stress through seed priming. Frontiers in Environmental Science. 2022;**10**:991977. DOI: 10.3389/fenvs.2022.991977

[24] Devika OS, Singh S, Sarkar D, Barnwal P, Suman J, Rakshit A. Seed priming: A potential supplement in integrated resource management under fragile intensive ecosystems. Frontiers in Sustainable Food Systems. 2021;**5**:654001. DOI: 10.3389/fsufs.2021.654001

[25] Banerjee A, Roychoudhury A. Seed priming technology in the amelioration of salinity stress in plants. In: Advances in Seed Priming. Singapore: Springer; 2018. pp. 81-93

[26] Farooq M, Basra SMA, Hafeez K. Seed invigoration by osmohardening in coarse and fine rice. Seed Science and Technology. 2006;**34**:181-187. DOI: 10.15258/sst.2006.34.1.19

[27] Alves R, Rossatto DR, da Silva J, Checchio MV, de Oliveira KR, de Oliveira AF, et al. Seed priming with ascorbic acid enhances salt tolerance in micro-tom tomato plants by modifying the antioxidant defense system components. Biocatalysis and Agricultural Biotechnology. 2021;**31**:101927

[28] González-García Y, López-Vargas ER, Pérez-Álvarez M, Cadenas-Pliego G, Benavides-Mendoza A, Valdés-Reyna J, et al. Seed priming with carbon nanomaterials improves the bioactive compounds of tomato plants under saline stress. Plants. 1984;**2022**(11):1984. DOI: 10.3390/plants11151984

[29] Alamer KH, Attia H. UV-C seed priming improves tomato plant growth against salt stress. Journal of Taibah University for Science. 2022;**16**(1):1181-1191

[30] Nawaz A, Amjad M, Jahangir MM, Khan SM, Cui H, Hu J. Induction of salt tolerance in tomato (*Lycopersicon esculentum* mill.) seeds through sand priming. Australian Journal of Crop Science. 2012;**6**(7):1199-1203

[31] Wu L, Huo W, Yao D, Li M. Effects of solid matrix priming (SMP) and salt stress on broccoli and cauliflower seed germination and early seedling growth. Scientia Horticulturae. 2019;**255**:161-168

[32] Hassini I, Martinez-Ballesta MC, Boughanmi N, Moreno DA, Carvajal M. Improvement of broccoli sprouts (*Brassica oleracea* L. var. italica) growth and quality by KCl seed priming and methyl jasmonate under salinity stress. Scientia Horticulturae. 2017;**226**:141-151

[33] Ghezal N, Rinez I, Sbai H, Saad I, Farooq M, Rinez A, et al. Improvement of *Pisum sativum* salt stress tolerance by bio-priming their seeds using Typha angustifolia leaves aqueous extract. South African Journal of Botany. 2016;**105**:240-250

[34] Yan M. Hydro-priming increases seed germination and early seedling growth in two cultivars of Napa cabbage (Brassica rapa subsp. *pekinensis*) grown under salt stress. The Journal of Horticultural Science and Biotechnology. 2016;**91**(4):421-426

[35] Borromeo I, Domenici F, Del Gallo M, Forni C. Role of polyamines in the response to salt stress of tomato. Plants. 2023;**12**(9):1855

[36] Roslan MAM, Zulkifli NN, Sobri ZM, Zuan ATK, Cheak SC, Abdul Rahman NA. Seed biopriming with P- and K-solubilizing Enterobacter hormaechei sp. improves the early vegetative growth and the P and K uptake of okra (Abelmoschus esculentus) seedling. PLoS One. 2020;**15**(7):e0232860. DOI: 10.1371/journal. pone.0232860

[37] Srivastava AK, Lokhande VH, Patade VY, Suprasanna P, Sjahril R, D'Souza SF. Comparative evaluation of hydro-, chemo-, and hormonal-priming methods for imparting salt and PEG stress tolerance in Indian mustard (*Brassica juncea* L.). Acta Physiologiae Plantarum. 2010;**32**:1135-1144

[38] Kaya MD, Okçu G, Atak M, Cıkılı Y, Kolsarıcı Ö. Seed treatments to overcome salt and drought stress during germination in sunflower (*Helianthus annuus* L.). European Journal of Agronomy. 2006;**24**(4):291-295

[39] Ghasemi N, Omidi H, Bostani A. Morphological properties of *Catharanthus roseus* L. seedlings affected by priming techniques under natural salinity stress. Journal of Plant Growth Regulation. 2021;**40**:550-557

[40] Tian Y, Guan B, Zhou D, Yu J, Li G, Lou Y. Responses of seed germination, seedling growth, and seed yield traits to seed pretreatment in maize (Zea mays L.). The Scientific World Journal. 2014;**2014**:834630

[41] Shukla N, Kuntal H, Shanker A, Sharma S. Hydro-priming methods for initiation of metabolic process and synchronization of germination in mung bean (Vigna radiata L.) seeds. Journal of Crop Science and Biotechnology. 2018;**21**:137-146

[42] Pedrini S, Balestrazzi A, Madsen MD, Bhalsing K, Hardegree SP, Dixon KW, et al. Seed enhancement: Getting seeds restoration-ready. Restoration Ecology. 2020;**28**:S266-S275

[43] Majda C, Khalid D, Aziz A, Rachid B, Badr AS, Lotfi A, et al. Nutri-priming

as an efficient means to improve the agronomic performance of molybdenum in common bean (Phaseolus vulgaris L.). Science of the Total Environment. 2019;**661**:654-663

[44] Nile SH, Thiruvengadam M, Wang Y, Samynathan R, Shariati MA, Rebezov M, et al. Nano-priming as emerging seed priming technology for sustainable agriculture—Recent developments and future perspectives. Journal of Nanobiotechnology. 2022;**20**(1):1-31

[45] Mahakham W, Sarmah AK, Maensiri S, Theerakulpisut P. Nanopriming technology for enhancing germination and starch metabolism of aged rice seeds using photosynthesized silver nanoparticles. Scientific Reports. 2017;**7**:8263

[46] Arun MN, Hebbar SS, Senthivel T, Nair AK, Padmavathi G, Pandey P, et al. Seed Priming: The Way Forward to Mitigate Abiotic Stress in Crops. Vol. 11. London, UK: IntechOpen; 2022. p. 173

[47] Alam MU, Fujita M, Nahar K, Rahman A, Anee TI, Masud AAKMR, et al. Seed priming upregulates antioxidant defense and glyoxalase systems to conferring simulated drought tolerance in wheat seedlings. Plant Stress. 2022;**6**:100120

[48] Varier A, Vari AK, Dadlani M. The subcellular basis of seed priming. Current Science. 2010;**99**:450-456

[49] Sano N, Seo M. Cell cycle inhibitors improve seed storability after priming treatments. Journal of Plant Research. 2019;**132**:263-271

[50] Sadak MS, Talaat IM. Attenuation of negative effects of saline stress in wheat plant by chitosan and calcium carbonate. Bulletin of the National Research Centre. 2021;**45**:1-12

[51] Chen X, Zhang R, Li B, Cui T, Liu C, Liu C, et al. Alleviation of oxidative damage induced by CaCl2 priming is related to osmotic and ion stress reduction rather than enhanced antioxidant capacity during germination under salt stress in sorghum. Frontiers in Plant Science. 2022;**13**:881039. DOI: 10.3389/fpls.2022.881039

[52] Ben Youssef R, Jelali N, Boukari N, Albacete A, Martinez C, Alfocea FP, et al. The efficiency of different priming agents for improving germination and early seedling growth of local Tunisian barley under salinity stress. Plants. 2021;**10**(11):2264

[53] Sairam RK. Effect of moisture-stress on physiological activities of two contrasting wheat genotypes. Indian Journal of Experimental Biology. 1994;**32**:594-594

[54] Yang Y, Guo Y. Elucidating the molecular mechanisms mediating plant salt-stress responses. New Phytologist. 2018;**217**(2):523-539

[55] Johnson R, Puthur JT. Seed priming as a cost effective technique for developing plants with cross tolerance to salinity stress. Plant Physiology and Biochemistry. 2021;**162**:247-257

[56] Bajwa AA, Farooq M, Nawaz A. Seed priming with sorghum extracts and benzyl aminopurine improves the tolerance against salt stress in wheat (*Triticum aestivum* L.). Physiology and Molecular Biology of Plants. 2018;**24**:239-249. DOI: 10.1007/s12298-018-0512-9

[57] Brocklehurst PA, Dearman J. Interaction between seed priming treatments and nine seed lots of carrot, celery and onion II. Seedling emergence and plant growth. The Annals of Applied Biology. 2008;**102**:583-593

[58] Bewley JD, Bradford K, Hilhorst H. Seeds: physiology of development, germination and dormancy. Berlin, Germany: Springer Science & Business Media; 2012

[59] Abdel-Aziz HMM, Hasaneen MNA, Omer AM. Impact of engineered nanomaterials either alone or loaded with NPK on growth and productivity of French bean plants: Seed priming vs foliar application. South African Journal of Botany. 2019;**125**:102-108

[60] Desta B, Amare G. Paclobutrazol as a plant growth regulator. Chemical and Biological Technologies in Agriculture. 2021;**8**:1. DOI: 10.1186/s40538-020-00199-z

[61] Duman I, Duzyaman E. Growth control in processing tomato seedlings. Acta Horticulturae. 2003;**613**:95-102

[62] Saidi I, Ammar S, Demont-Caulet N, Thévenin J, Lapierre C, Bouzid S, et al. Thigmomorphogenesis in *Solanum lycopersicum*: Morphological and biochemical responses in stem after mechanical stimulation. Plant Science. 2009;**177**:1-6

[63] Sabina S, Jithesh MN. Mechanical wounding of leaf midrib and lamina elicits differential biochemical response and mitigates salinity induced damage in tomato. Journal of Applied Horticultre. 2021;**23**:3-10

[64] Depège N, Thonat C, Coutand C, Julien JL, Boyer N. Morphological responses and molecular modifications in tomato plants after mechanical stimulation. Plant & Cell Physiology. 1997;**38**(10):1127-1134. DOI: 10.1093/oxfordjournals.pcp.a029097

[65] Garner LC, Björkman T. Mechanical conditioning of tomato seedlings improves transplant quality without deleterious effects on field performance. Hort Science. 1999;**34**:848-851

[66] Van Gaal T, Erwin JE. Diurnal variation in Thigmotropic inhibition of stem elongation. Hort Technology. 2005;**15**:291-294

[67] Sparke M, Wegscheider A, Winterhagen P, Ruttensperger U, Hegele M, Wuensche J. Air-based mechanical stimulation controls plant height of ornamental plants and vegetable crops under greenhouse conditions. Hort Technology. 2021;**31**:405-416

[68] Li Y, Li J, Wu G, Li Y, Shen A, Ma D, et al. Design of an air blowing device above seedbed: The effect of air disturbance on the microenvironment and growth of tomato seedlings. HortScience. 2020;**55**:1308-1314

[69] Uchida A, Yamamoto KT. Effects of mechanical vibration on seed germination of Arabidopsis thaliana (L.) Heynh. Plant and Cell Physiology. 2002;**43**:647-651

[70] Yang JL, Dong HZ, Hou QJ, Li Q. Effect of ultrasonic waves on seed germination ability of two kinds of vegetables. Seed. 2011;**30**(12):97-e98

[71] Capiati DA, Pais SM, Tellez-Iñon MT. Wounding increases salt tolerance in tomato plants: Evidence on the participation of calmodulin-like activities in cross-tolerance signaling. Journal of Experimental Botany. 2006;**57**:2391-2400

[72] Benikhlef L, L'Haridon F, Abou-Mansour E, Serrano M, Binda M, Costa A, et al. Perception of soft mechanical stress in Arabidopsis leaves activates disease resistance. BMC Plant Biology. 2013;**13**:133. DOI: 10.1186/1471-2229-13-133

[73] Keller E, Steffen KL. Increased chilling tolerance and altered carbon metabolism in tomato leaves following application of mechanical stress. Physiologia Plantarum. 1995;**93**(3):519-525

[74] Si T, Wang X, Zhao C, Huang M, Cai J, Zhou Q, et al. The role of hydrogen peroxide in mediating the mechanical wounding-induced freezing tolerance in wheat. Frontiers in Plant Science. 2018;**9**:1-15

[75] Li X et al. Mechano-stimulated modifications in the chloroplast antioxidant system and proteome changes are associated with cold response in wheat. BMC Plant Biology. 2015;**15**:219. DOI: 10.1186/s12870-015-0610-6

[76] Gong M, Chen BO, Li ZG, Guo LH. Heat-shock-induced cross adaptation to heat, chilling, drought and salt stress in maize seedlings and involvement of H_2O_2. Journal of Plant Physiology. 2001;**158**(9):1125-1130

[77] López-Ribera I, Vicient CM. Drought tolerance induced by sound in *Arabidopsis* plants. Plant Signaling & Behavior. 2017;**12**(10):e1368938

[78] Ghalkhani E, Hassanpour H, Niknam V. Sinusoidal vibration alleviates salt stress by induction of antioxidative enzymes and anatomical changes in *Mentha pulegium* (L.). Acta Physiologiae Plantarum. 2020;**42**:1-13

[79] Suge H. Dehydration and drought resistance in *Phaseolus vulgaris* as affected by mechanical stress. Reports of the Institute for Agricultural Research, Tohoku University. 1980;**31**:1-10

[80] Dombrowski JE. Salt stress activation of wound-related genes in tomato plants. Plant Physiology. 2003;**132**(4):2098-2107

[81] Sic Zlabur J et al. Plant response to mechanically-induced stress: A case study on specialized metabolites of leafy vegetables. Plants. 2021;**10**:2650. DOI: 10.3390/plants10122650

[82] Mierziak J, Kostyn K, Kulma A. Flavonoids as important molecules of plant interactions with the environment. Molecules. 2014;**19**:16240-16265

[83] Hefferon KL. Nutritionally enhanced food crops; progress and perspectives. International Journal of Molecular Sciences. 2015;**16**(2):3895-3914. DOI: 10.3390/ijms16023895

[84] Howell AH, Völkner C, McGreevy P, et al. Pavement cells distinguish touch from letting go. Nature Plants. 2023;**9**(6):877-882. DOI: 10.1038/s41477-023-01418-9

[85] Erner Y, Biro R, Jaffe MJ. Thigmomorphogenesis: Evidence for a translocatable thigmomorphogenetic factor induced by mechanical perturbation of beans (*Phaseolus vulgaris*). Physiologia Plantarum. 1980;**50**:21-25

[86] Börnke F, Rocksch T. Thigmomorphogenesis—Control of plant growth by mechanical stimulation. Scientia Horticulturae (Amst.). Netherlands: Elsevier; 2018;**234**:344-353

[87] Braam J, Davis RW. Rain-, wind-, and touch-induced expression of calmodulin and calmodulin-related genes in Arabidopsis. Cell. 1990;**60**:357-364

[88] Xu W, Purugganan MM, Polisensky DH, Antosiewicz DM, Fry SC, Braam J. Arabidopsis TCH4, regulated by hormones and the environment, encodes a xyloglucan endotransglycosylase. The Plant Cell. 1995;**7**(10):1555-1567

[89] Lee D, Polisensky DH, Braam J. Genome-wide identification of touch- and darkness-regulated Arabidopsis genes: A focus on calmodulin-like and XTH genes. The New Phytologist. 2005;**165**:429-444

[90] Kurusu T, Kuchitsu K, Nakano M, Nakayama Y, Iida H. Plant mechanosensing and Ca2+ transport. Trends in Plant Science. 2013;**18**: 227-233

[91] Hamilton ES, Schlegel AM, Haswell ES. United in diversity: Mechanosensitive ion channels in plants. Annual Review of Plant Biology. 2015;**66**:113-137

[92] He Y, Zhou J, Shan L, Meng X. Plant cell surface receptor-mediated signaling – A common theme amid diversity. Journal of Cell Science. 2018;**131**:jcs209353

[93] Ghosh R, Barbacci A, Leblanc-Fournier N. Mechanostimulation: A promising alternative for sustainable agriculture practices. Journal of Experimental Botany. 2021;**72**(8):2877-2888. DOI: 10.1093/jxb/erab036

[94] Schnelle MA, McCraw BD, Schmoll TJ. A brushing apparatus for height control of bedding plants. Hort Technology. 1994;**4**:275-276

[95] Koch R, Sauer H, Ruttensperger U. Einfluss von mechanischen Berührungsreizen auf das Wachstum von Küchenkräutern im Topf. Gesunde Pflanzen. 2011;**63**:199-204

[96] Richter E. Nützlingseinsatz im Zierpflanzenbau unter Glas: Handbuch zum praktischen Arbeiten: Ergebnisse aus den Verbundvorhaben "Nützlinge I" und "Nützlinge II" DPG Selbstverlag. Braunschweig: DPG; 2009

[97] Rocksch T, Helbig D, Grüneberg H. Alternative Wachstumsregulierung von Pflanzen mittels gesteuerter Vibrationstechnik - Versuchsaufbau und Erfassung technischer Parameter. In: Gartenbauwissenschaftliche Jahrestagung DGG & BHGL. Vol. 48. Bonn, Berlin. 2013. p. 32

[98] Grüneberg H, Helbig D, Rocksch T. Alternative Wachstumsregulierung von Pflanzen mittels gesteuerter Vibrationstechnik als Ersatz chemischer Hemmstoffe. In: Schlussbericht BÖLN-Projekt, FKZ 2811NA010. Berlin. 2016. p. 57

[99] Kurtser P, Hanell U, Andreasson H. Robotic platform for precise mechanical stress induction in greenhouses cultivation. In: IEEE 16th International Conference on Automation Science and Engineering (CASE), Hong Kong, China, 2020. 2020. pp. 1558-1565. DOI: 10.1109/CASE48305.2020.9249229

[100] Kurtser P, Castro Alves V, Arunachalam A, Sjöberg V, Hanell U, Hyötyläinen T, et al. Development of novel robotic platforms for mechanical stress induction, and their effects on plant morphology, elements, and metabolism. Scientific Reports. 2021;**11**:23876. DOI: 10.1038/s41598-021-02581-9

[101] Wang K, Law K, Leung M, Wong W, Li NA. A labor-saving and repeatable touch-force signaling mutant screen protocol for the study of Thigmomorphogenesis of a model plant Arabidopsis thaliana. Journal of Visualized Experiments. 2019;**151**:e59392. DOI: 10.3791/59392

[102] Paparella S, Araújo SS, Rossi G, Wijayasinghe M, Carbonera D, Balestrazzi A. Seed priming: State of the art and new perspectives. Plant Cell Reports. 2015;**34**:1281-1293

[103] Karim MN, Sani MNH, Uddain J, Azad MOK, Kabir MS, Rahman MS, et al. Stimulatory effect of seed priming as pretreatment factors on germination and yield performance of yard long bean (Vigna unguiculata). Horticulturae. 2020;**6**(4):104

[104] Iida H. Mugifumi, a beneficial farm work of adding mechanical stress by treading to wheat and barley seedlings. Frontiers in Plant Science. 2014;**5**:453. DOI: 10.3389/fpls.2014.00453

Chapter 6

The Use of Static Magnetic Field on Irrigation Water and Its Effect on Mineral and Nutrient Content in *Solanum lycopersicum* L.

Albys Esther Ferrer Dubois, Yilan Fung Boix, Clara Martínez Manrique, Liliana Gómez Luna, Elizabeth Isaac Aleman, Sophie Hendrix, Natalie Beenaerts and Ann Cuypers

Abstract

Solanum lycopersicum L. is a nutraceutical plant. Tomato yield and morphological features have been improved by irrigation with water treated with static magnetic field (SMF). The effect of magnetically treated water with SMF (20–200 mT) on mineral and nutritional contents in *Solanum lycopersicum* L. was investigated. Bromatological analyses and minerals contents were determined in tomato fruits harvested after the application of two irrigation protocols (water treated with SMF between 20 and 200 mT and water without SMF treatment as control). Fruits were selected for analysis according to a completely randomized design. Although no significant differences were observed between both groups with regard to bromatological analyses (moisture, total ash, total solids, proteins), an increasing trend was determined for these components in fruits of plants irrigated with SMF-treated water. An increase was detected for potassium (K), calcium (Ca) and cupper (Cu) concentrations in these conditions as compared to fruits of control plants. The SMF treatment of irrigation water improves the nutrient uptake and the water use efficiency in tomato. The nutraceutical value of these fruits was increase and can be considered as an important strategy for future crop production to improve food quality.

Keywords: *Solanum lycopersicum* L., static magnetic field, minerals, nutritional ingredients, fruits

1. Introduction

Tomato (*Solanum lycopersicum* L.) fruits are known to have some therapeutic activity in the prevention and treatment of cancer, cardiovascular and degenerative diseases [1–4]. This plant is considered as a nutraceutical plant due to its antioxidant properties determined by its content in carotenoids, polyphenols, organic acids and vitamins, exerting an effective action in metabolic processes in the human diet and nutrition [5–10]. Together with antioxidant enzymes and secondary metabolites, they participate in the defense of organisms against oxidative stress, which, when not controlled, might lead to damage. In addition, this plant has high nutritional value because of its mineral content in its fruits [11, 12].

Solanum lycopersicum L. is often used as a biological model organism in plant growth studies, since results obtained in tomato are also applicable to other horticultural species [13, 14]. As its genome is entirely sequenced, the tomato is a very useful model organism for genetic analysis of many aspects of reproductive plant development. In addition, it permits the study of growth patterns, body architecture, and fruit ripening, all of great agronomic interest [15, 16].

In previous studies, it has been shown that tomato yield and morphology are improved by irrigation with water treated with static magnetic field (SMF) [17–26]. The irrigation protocol using SMF-treated water (20–200 mT) increased the size of the fruits and vegetative organs of tomato plants [27–30]. Moreover, it has been demonstrated that the use of electromagnetic fields (EMF) can activate cellular functions and improve the yield [31–37]. The synthesis of bioactive compounds like carotenoids, lycopene, polyphenols, ascorbic acid and quercetin can be increased, as well as the antioxidant capacity in tomato juice [38–42].

Despite these promising results, there is currently no evidence for the effects of irrigation with SMF-treated water on the mineral content of *Solanum lycopersicum* L., limiting its applications. In the current study, the concentrations of macro- and micronutrients [calcium (Ca), magnesium (Mg), phosphorus (P), potassium (K), sulfur (S), copper (Cu), manganese (Mn), sodium (Na), zinc (Zn)] were determined in tomato fruits. In addition, to evaluate possible contamination via irrigation water, also the concentrations of the non-essential elements cadmium (Cd) and lead (Pb) were analyzed. Complementary to nutrient values, fruit quality parameters such as protein content, and moisture content as well as total soluble solids and ashes were compared between fruits of tomato plants irrigated with SMF-treated water and fruits of control plants.

2. Materials and methods

This research was conducted in the National Center of Applied Electromagnetism (CNEA, University of Oriente, Cuba), in the greenhouse "Campo Antena" (Santiago of Cuba), and Environmental Sciences Laboratory (Hasselt University, Belgium).

2.1 Plant growth, irrigation and harvest

2.1.1 Plant material and growth conditions

The tomato species used was *Solanum lycopersicum* L. hybrid Aegean, identify by BSC 21509, in BIOECO (Oriental Center of Ecosystems and Biodiversity, Santiago de Cuba, Cuba). The Provincial Laboratory of Seeds, which belongs to the Ministry of Agriculture (MINAG) in Santiago de Cuba, provided the certified seeds. Plants were cultivated under semi-controlled conditions in a greenhouse Type I, Model G-2 MCC during 4 months. The substrate composition used met the

Indicative	Properties	Soil
Physics	Soil type	Sour igneous rock
	Texture	Clay with lightly wavy slope
	Depth of superficial soil	25–50 cm
	Density	1.0–1.3 g cm^{-3}
	Erosion	Limited
	Infiltration	15–20%
	Water retention capacity (saturation)	Saturated, medium humidity (420 cm)
	Type	Brown without carbonate
Chemical	Organic matter	3.5%
	pH	7.0
	Electrical conductivity	1.39–2.01(mS cm^{-1})
	Phosphorus	7.3 meq100 g of $soil^{-1}$
	Calcium	38.7 meq100 g of $soil^{-1}$
	Potassium	21.7 meq100 g of $soil^{-1}$

Table 1.
Soil quality parameters for the cultivation of Solanum lycopersicum L. in the Campo Antena greenhouse.

requirements of the Agriculture Ministry of Cuba with regard to the content of organic matter, ions, pH, conductivity and microbiological composition (**Table 1**). The temperature in the experimental growth area (Campo Antena, a suburban area of Santiago de Cuba) ranged from 27 to 30°C. The approximate value of precipitation was 68.04 mm and the relative humidity 70.32% during the entire period of the experiment.

Using the established protocols [43], soil and water quality were analyzed prior to the application of the irrigation protocol. The physical and chemical parameters of the soil quality were determined according to protocols [44, 45]. Three soil samples were randomly chosen at several places and mixed to obtain one experimental sample. Microbiological analysis was performed according to the requirements of the Ministry of Agriculture of Cuba [46].

2.1.2 Irrigation system and magnet device

For the irrigation, an air microjet device was used and the irrigation system was equipped with a pump characterized by a pump flow of 2.54–2.91 $m^3 h^{-1}$ and a speed ranging from 1.4 to 1.6 m s^{-1}. The timing of irrigation was based on the cultivation phase of the plants. During the first and second cultivation phase (i.e. establishment, vegetative growth and flowering), the irrigation was performed once a day for 20 min. In the third cultivation phase, which incorporates fruit ripening, the irrigation was conducted twice a day during 20 min. The irrigation schedule was completed after 120 days.

The magnetic treatment was applied used a magnetic device composed of permanent magnets with a non-uniform SMF. This was designed, constructed and characterized in CNEA [47]. The treatment was conducted at the time of irrigation of the plants during the whole cycle of the crop growth.

2.1.3 Experimental design

For the application of the SMF treatment, a totally randomized experimental design with three replicates per experimental group (control plants or plants irrigated with SMF-treated water) was used. The experiment consisted of 25 plants per treatment making up 150 plants. After 120 days, all fruits were harvested. Ten fruits per group were chosen randomly to perform the analyses.

The experimental groups were described as follows:

- Control plants: *Solanum lycopersicum* L. irrigated with water not treated withSMF.
- SMF plants: *Solanum lycopersicum* L. irrigated with water treated with SMF between 20 and 200 mT.

The selection of the induction intensity was based on the recommendation provided in literature. In *Solanum lycopersicum* L., magnetic inductions ranging between 120 and 250 mT were used to induce seed germination, enhance the morphological features and increase the average yield per plant [16, 21, 27].

2.2 Bromatological analyses

Fruit analyses were performed on fruits harvested between the fourth (fruit ripening) and 6th stage (senescence) according to the technical protocols recommended, because a similar maturation in the majority of the fruits was obtained in this stage [48]. In the bromatological analyses, 10 mature fruits from each experimental group were randomly selected and the analyses were performed in triplicate.

2.2.1 Determination of moisture content

The determination of the moisture content was performed according to the gravimetric method described for fruits in the Official Methods of Analysis [48]. Approximately 0.5 g of ripe fruits of *Solanum lycopersicum* L. (fresh weight), were oven-dried at 60°C for 72 h until a constant weight (dry weight) was reached. The moisture content was calculated.

2.2.2 Determination of total ashes content

The total ashes content was determined by the gravimetric method [49]. Fruit samples (10 g) from *Solanum lycopersicum* L. were placed in apre-weighed silica crucible. The crucible was heated at 100°C until the material was completely charred. Thereafter, it was heated to 600°C in a muffle furnace for 3–5 h, cooled in a desiccator and weighed. In order to ensure complete incineration, the crucible was re-heated in the oven for 30 min, cooled and weighed again. This process was repeated until the ash weight is constant. The total ashes content was calculated.

2.2.3 Determination of the protein content

The determination of the total soluble proteins in alkaline medium was performed by the bicinchoninic acid (BCA) method [50]. Dried fruits (10 mg) from

Solanum lycopersicum L. were placed in a centrifuge tube and 1 mL of NaOH (0.5 M) was added to each sample for alkaline digestion. The samples were placed in a water bath at 100°C for 10 min. Subsequently, they were centrifuged at room temperature (13,000 × *g*) for 5 min. BCA reagent (AppliChem, Germany) (1 mL) was added to 100 μL of the supernatant. The samples were mixed and after 15 min the absorbance was measured in a UV-visible spectrophotometer at 562 nm. A standard curve using bovine serum albumin (BSA) (15–125 μg mL^{-1}) was used. The results were expressed as total protein concentration in μg protein per gram of dried weight.

2.2.4 Determination of total soluble solid content

For the determination of the total soluble solid content, a gravimetric method was used [51]. Three porcelain capsules were dried in an oven at 105°C for 60 min. The capsules were placed in a desiccator until they were completely cooled and then weighed. Aqueous extracts (50 mg mL^{-1}) of dried fruits from *Solanum lycopersicum* L. were prepared in distilled water. One mL of this fruit extract was added to each capsule. The filled capsules were weighed and placed in an oven at 105°C for 60 min. After this, they were placed in a desiccator until they were completely cooled and then weighed again. The drying process was repeated until a constant weight was obtained. The percentage of total solids was calculated.

2.3 Determination of mineral elements

The concentrations of mineral elements were determined after acidic extraction of the fruits using a heating block. Five ripe fruits per experimental group were dried in an oven for 72 h at 60°C. The extraction of the elements was carried out according to the procedures of the Ministry of Public Health of Cuba. The dried fruits from *Solanum lycopersicum* L. were weighed, kept at 4°C and pulverized to powder in an electric mill. For an optimal extraction of the elements, this process was conducted away from moisture and light according to the requirements. The powder (50 mg) was digested using a mixture (1:1, v/v) of concentrated nitric acid (70%) and hydrochloric acid (37%). The resulting material was filtered.

The macroelements and microelements were determined in the extracts of fruits of *Solanum lycopersicum* L. using inductively coupled plasma emission spectroscopy (ICP-OES) (Series, Agilent Technologies, Dieghem, Belgium). The analysis was validated using *Solanum lycopersicum* L. certified as reference material (CRM 1573a), from the National Institute of Standards and Technology [52]. HNO_3 solutions (1%) were used as blank. Calibration curves were generated to identify each mineral.

2.4 Statistical analysis

For statistical analysis, normal distribution of the data was tested by the Kolmogorov-Smirnov test. Parametric tests were applied to compare the results of the experimental groups. Either the Student t-Test or one-way Analysis of Variance (one-way ANOVA) followed by the post-hoc Tukey-Kramer test were performed. The value of statistical significance was set at 95%.

3. Results

3.1 Soil properties

To establish the cultivation of *Solanum lycopersicum* L. under semi-controlled experimental conditions, it is necessary to know the quality of the soil and irrigation water. In **Table 1**, the soil quality parameters are shown according to the classification of soils in Cuba for the tomato cultivation [45]. The typology of the soil is in agreement with the established quality parameters of the Ministry of Agriculture and reached the required nutritional levels. In addition, neither parasites nor nematodes were detected in the microbiological analysis.

3.2 Irrigation water properties

Water has several important basic functions in plants. It is the biggest constituent of the cytoplasm and vacuole providing turgor (85–95%). Furthermore, it is required for dissolving and transporting nutrients. Finally, it provides the electrons necessary for the redox reactions involved in photosynthesis [45]. When water is exposed to SMF with a certain induction and speed, it acquires different physicochemical characteristics (**Table 2**) [53, 54].

3.3 Bromatological analysis

To investigate whether alterations in irrigation water also can affect the nutritional quality of the tomato fruits, nutritionals indicators and mineral concentrations were determined in fruits of plants grown using irrigation water that was either treated with SMF in a range of 20–200 mT or where no treatment was applied. From here on, the group of plants irrigated with SMF-treated water is referred to as SMF plants, whereas the plants irrigated with water without the SMF treatment are referred to as

Characteristic	Irrigation water	
	Control	SMF-treated
Electrical conductivity (mS cm^{-1})	0.019 ± 0.00	0.25 ± 0.02*
pH	7.14 ± 0.02	7.87 ± 0.02
K (mg L^{-1})	4.17 ± 0.84	5.39 ± 0.81
Ca (mg L^{-1})	24.7 ± 0.90	24.53 ± 1.27
Na (mg L^{-1})	11.47 ± 0.95	12.17 ± 0.78
Mg (mg L^{-1})	6.02 ± 1.00	6.39 ± 0.91
SO_4 (mg L^{-1})	9.24 ± 0.48	9.25 ± 0.27
CO_3 (mg L^{-1})	17.33 ± 1.15	17.33 ± 0.58
Cl (mg L^{-1})	15.8 ± 1.8	17.31 ± 1.12
HCO_3 (mg L^{-1})	72.88 ± 0.95	73.41 ± 0.99

*Notes: Values represent the mean ± SE (n = 3). *Indicates a significant difference between both treatment groups ($p \leq 0.05$) Student's t-test).*

Table 2.
Characteristics of the irrigation water used for the cultivation of Solanum lycopersicum L.

Indicators	Experimental groups	
	Control plants	SMF plants
Moisture (%)	93.26 ± 1.30	95.23 ± 0.77
Ashes (%)	0.58 ± 0.04	0.64 ± 0.01
Total proteins (μg mL^{-1})	53.35 ± 17.32	59.85 ± 27.93
Total solids (%)	7.20 ± 1.88	7.63 ± 2.06

Notes: Control plants: Solanum lycopersicum L. irrigated with non-SMF-treated water; SMF plants: Solanum lycopersicum L. irrigated with SMF-treated water (20–200 mT). Data represent the mean ± SE (n = 3).

Table 3.
Analysis of nutritional indicators in Solanum lycopersicum L. fruits irrigated with water subjected to different treatments.

Minerals (mg kg^{-1} DW)	Experimental groups	
	Control plants	SMF plants
Potassium (K)	25.02 ± 0.42	28.58 ± 1.12*
Calcium (Ca)	1.30 ± 0.14	1.77 ± 0.05*
Sodium (Na)	0.37 ± 0.01	0.42 ± 0.02
Magnesium (Mg)	1.21 ± 0.10	1.22 ± 0.05
Copper (Cu)	0.01 ± 0.00	0.23 ± 0.04*
Phosphorous (P)	3.03 ± 0.03	3.35 ± 0.01
Sulfur (S)	1.45 ± 0.06	1.44 ± 0.06

*Notes: Control plants: Solanum lycopersicum L. irrigated with non-SMF-treated water; SMF plants: Solanum lycopersicum L. irrigated with SMF-treated water (20–200 mT). Data represent the mean ± SE (n = 3). *Symbols indicate statistically significant differences between both experimental groups Student's t-test, ($p < 0.05$).*

Table 4.
Mineral content in fruits of Solanum lycopersicum L. irrigated with water subjected to different treatments.

control plants. The results of the bromatological analyses of *Solanum lycopersicum* L. SMF plants and control plants are shown in **Table 3**.

3.4 Mineral elements analysis

In order to further determine the nutritional value of tomato plants, the concentrations of seven different elements essential for plant growth and development were determined in the fruits of *Solanum lycopersicum* L. (**Table 4**).

4. Discussion

4.1 Soil properties

The water retention capacity estimates the availability of water to plants that is essential for normal growth. This parameter is among others correlated with the organic matter content. The organic matter content determined in this experiment has an average value similar to the established range between 1.65 and 2.65% [55].

According to reports, the electrical conductivity for the growth of *Solanum lycopersicum* L. should be between 0.75 and 2.5 mS cm^{-1} [12]. This coincided with the results obtained from the soil analysis of our experimental setup.

K and Ca are two primary minerals in the soil that were determined. Ca salts are predominantly present, contributing to the classification of the soil as neutral, with a pH close to 7. In this soil, no nutrients reached toxic levels according to the classification found in literature [45]. These authors considered that a neutral soil saturated with basic ions (K^+, Ca^{2+}, Mg^{2+}, Na^+) lose these minerals less than a soil with a low percentage of saturation [45]. Therefore, it can be supposed that plants easily assimilate these minerals. The chemical indicators provide knowledge on the nutritional state of the soil, associated with its fertility. The results of the soil analysis showed that the nutrient levels were adequate for the growth of *Solanum lycopersicum* L.

4.2 Irrigation water properties

After applying SMF, an increase in the electrical conductivity with 0.23 units was observed and therefore it has a bigger capacity to drive an electric current (**Table 2**). Due to its electrical conductivity and the presence of Na, the salinity of the water is classified as low to moderate according to its use for irrigation purposes [56]. The increase of the electrical conductivity could be influenced by an increase in the mobility of the ions in the water treated with SMF during irrigation of the *Solanum lycopersicum* L. plants. This may be due to weakening of the hydration shell around the H^+ and OH^- ions [57].

The effects of MF treatment on the electrical conductivity of water are controversial. The studies of Martínez et al. [58], support the results of this research, because they reported an increase of 10.75–12 mS cm^{-1} in the ionic conductivity for aqueous solutions of salts of sodium and chloride, to a concentration 0.1% in water treated with SMF in the range from 10 to 160 mT. These authors explained that these variations depend on the movement of the loaded ions. This are related with the changes in the distribution of the size of the polymeric species of the aqueous solutions. An increase in conductivity could be due to changes in the polarity of the water molecules, as a consequence of the changes at the dipole moment, in the electron and vibrational transition state in the molecule. All that which modifies the properties of the water and it favors the mobility of the ions for a bigger hydrate. The treatment with small magnetic inductions (smaller than 1 T), can modify the organization of the molecules of water of big to small groups, each one of them composed symmetrically by six molecules [54]. Different variations in the microscopic structures of water under the action of the MF, with an increase in their electrical conductivity have been described. Similar changes were reported with SMF of 150 mT [59]. The irrigation water through a MF with 100 mT increased the electrical conductivity [60]. In plants of *Zea mays* an increase of electrical conductivity after a magnetic treatment in the irrigation water of 1500 mT was obtained [61].

The soil pH of the irrigation water was in the neutral range (pH 7) (**Table 2**). However, an increasing trend in pH of SMF-treated irrigation water was observed. Also described differences in water treated with SMF to favors the formation of $Ca(CO_3)_2$ and other alkaline materials that slightly reduced the acidity of the water [53]. Results of the water analysis indicated that there was a slight increment in ionic concentrations of K, Na, Cl in the SMF-treated water as compared to the control water, although no significant differences were observed.

Another important aspect of the water quality to be considered is the hardness of the water. The most common minerals defining water hardness are carbonates and sulphates of Mg and Ca. Water with a total hardness in which mineral concentrations are less than 27 ppm is categorized as soft. Moderately hard water has mineral concentrations in the range of 60–120 ppm of these elements and very hard water exceeds 180 ppm [62]. Taking these values into account, under the conditions of our experiment in which Ca and Mg concentrations were 24 and 6 ppm, respectively (**Table 2**), we can concluded that the water used was categorized as soft water.

The main components of soluble salts of irrigation water are Ca^{2+}, Mg^{2+}, Na^{+}, and K^{+} minerals has been demonstrated [63]. Ca^{2+} and Mg^{2+} ions cause hardness of water, in addition to dissolved metals, carbonates (CO_3^{2-}), bicarbonates (HCO^{3-}), chlorides (Cl^-) and sulfates (SO_4^{2-}) as dominant soluble salts. In general, magnetic treatment of water is thought to modify its structure and hence the dissolution of the minerals it contains [57]. Several studies explain about the theory of magnetization of water. According to the researches, the SMF causes physical and chemical changes in the distribution and polarization of the microscopic structure of molecules of water [64].

Experimental results have been described of infrared spectroscopy analysis of magnetically treated of water [65]. Theses confirm the variations in the physicochemical properties of the SMF-treated water in our experimental setup. In general, the surface tension was reduced and viscosity and vaporization rates were increased. These three phenomena, just as the transference of protons in the bonds by hydrogen bridges, are consequences of the modification of the molecular water structure [66]. The SMF treatment might have caused changes in the water molecule size due to the extra hydrogen bonds formation that is relative to the exposure time. The magnetic force is affecting the water molecule and disturbs dehydration phenomena by changing the orientation of the molecules. The hydrogen bonds between the molecules either changes or are released and this might release sucking energy and decrease the unity of water parts [67]. Others authors also report that MF modified the chemical structure of the water to be more clustered together [68]. The magnetically treated water can arrange its molecules in one direction. Again, the changes in the direction of these molecules might lead to changing order composing hydrogen bonds between molecules [69]. Overall, magnetic treatment of irrigation water affects the molecular structure of the water, whereas the molecular composition remains the same [53].

Water is a bipolar molecule and it allows electrostatic attraction and hydration of ions. One of its functions in plants is to act as a solvent for minerals [24]. Magnetic treatment improves the dissolving properties of water, which might result in an increased ability of nutrient assimilation by plants. Several researchers have verified that after applying MF to water, changes in the hydration of the ions occur and as a consequence, the water molecules bound more easily to soil particles, penetrating the micropores of the soil and preventing migration to greater depths [57, 66, 70]. Under conditions of this experiment, the results suggest that application of irrigation with SMF improved penetration of water in the soil and solubilization of nutrients. This is important for a better use of water by plants, which can contribute to increase growth.

4.3 Bromatological analysis

Although no statistical significant differences could be observed for the different parameters analyzed in *Solanum lycopersicum* L. fruits, a slight increment in total protein content (12.18%), ashes (10.34%), total solids (7.16%) and moisture (2.15%)

was observed in the SMF plants as compared to the control plants. In general, this points toward an overall improvement of the nutritional value of the SMF plants.

An increase in the total protein content is considered beneficial for plants. This might be linked to the dynamic properties of the cell wall, which are modified during growth and differentiation. In the cell wall, proteins containing glycine, proline and hydroxyproline are abundant. They are related to cell wall resistance and protection against pathogens and the reactive oxygen species (ROS) production [71]. Although irrigation of *Solanum lycopersicum* L. plants with SMF-treated water did not significantly increased the protein content, an increasing trend could be determined (**Table 3**). Increased protein biosynthesis was also indicated in other studies. A significant increase in the content of total proteins and proline after irrigation of *Vicia faba* L. with water treated with a MF was demonstrated [72]. Similar results were obtained in *Triticum sp.*, *Linum bienne*, *Cicer arietinum* and *Lens culinaris* plants irrigated with magnetically treated water [73].

Concerning the percentage of ashes found in *Solanum lycopersicum* L. fruits (**Table 3**), a slight increment was observed in fruits of SMF plants, as compared to the content of control plants. The effects of irrigation with SMF-treated water on the plants can explain this behavior. The SMF-treated water is better assimilated or absorbed by plant cells and can affect the membrane potential was suggested [74]. Therefore, the plant uptake of water and minerals was favored and it was faster than in normal conditions, due to the activation of the osmosis mechanism. Similar effects in the increased uptake of nutrients were reported [75]. They showed significant effects of non-uniform SMF of 50.6, 108.7 and 332 mT on the water absorption of *Solanum lycopersicum* L. seeds, possibly leading to increase moisture content in the fruits. The ashes percentage in tomato fruits in a ranged between 0.56 and 0.70% was described [76]. These are in agreement with the results obtained in this research. The ashes are the residue of the complete combustion of the organic material of the fruits and represent the mineral fraction. Therefore, the ashes content predicts the presence of minerals, especially K, Ca, S and Mg [77].

In the context of these observations, there are reports describing an increase in the absorption and assimilation of nutrients in *Lens culinaris* plants irrigated with magnetically treated water [78]. Positive effects were also obtained under these conditions in *Vigna unguiculata* a result of an increased absorption and assimilation of nutrients [79]. The moisture level is an indicator of the degree of hydration of tissues and an important determinant of fruit quality. Several studies reported that tomato fruits have high water content, ranging between 90 and 93.8% [80, 81]. These data are in agreement with those obtained in the control plants in the present research. Results show that fruits of SMF plants had slightly higher moisture content as compared to the control plants, suggesting a more efficient water absorption (**Table 3**).

The irrigation of some plant species with SMF-treated water activates cellular functions, thereby affecting physiological and biochemical processes, leading to increased primary metabolite levels was reported [82]. Among them are amino acids, lipids, nucleotides and proteins, which are used in combination with water and minerals in different plant processes including photosynthesis, respiration and nutrient transport and assimilation. The levels of these primary metabolites are related to the total solids levels, because they constitute the dry matter remaining after the removal of water. The total solids levels of the fruits were similar between both experimental groups and are in the established range for *Solanum lycopersicum* L. plants although a slight increase was observed in fruits of SMF plants (**Table 3**). Under normal growth conditions, the content of total solids in this plant was in a range between 3.5 and

7 ^{0}Brix was indicated [83, 84]. Whereas we did not observe a significant difference between both experimental groups, some authors explained that irrigation with water treated with a 50 Hz of EMF, can have an effect on the total solid content between 4 and 7% increase in the same variety of plant [85]. Water is essential for metabolic and physiological processes in living organisms for transporting nutrients and minerals necessary for fruit ripening. During this process, the content of total solids generally increases, which influences the biochemical quality of the fruits, together with the sugars and the total acidity, determining their flavor [83].

The results obtained in this research are also in good agreement with others researchers. The magnetically treated water had a positive effect on the nutritional qualities of *Citrullus lanatus* [86]. Theses authors founded a higher percentage of water content, crude protein and ashes content in plants irrigated by magnetically treated water when compared to the corresponding values of control plants. The relevance of the results obtained in the stimulation of ashes and proteins content in fruits of *Solanum lycopersicum* L. after magnetic treatment in irrigation water is significant. This is due to the fact that these nutritional indicators are determinant in the bromatological analyzes of the plants used for human consumption.

4.4 Mineral elements analysis

The positive effects detected in the nutrient concentrations of fruits in this research, indicate a better assimilation of nutrients and fertilizers by plants when they are irrigated with SMF-treated water during their growing cycle (**Table 4**). These findings are agreement with results revealed [87]. In general, it is clear that fruits of the SMF plants contained significantly higher concentrations of K, Ca and Cu as compared to the control group ($p < 0.05$).

The macronutrients like K, P, Ca, Mg and S have an important nutritional value in tomato fruits [45]. The higher mineral levels were found in the fruits of the SMF plants (**Table 4**). These results were in agreement with the increased ashes percentage obtained. Therefore, this parameter has a predictive value for nutritional value related to elemental uptake as previously indicated (**Table 3**).

By far, K exhibited the highest concentration of minerals determined. Similar results have been described when 85 genotypes of tomatoes were analyzed [88]. K plays an important role in the quantity of sugars that are accumulated in the fruit. It helps to increase the quantity of dry matter and vitamin C [12]. The same applies to Na, which although being a micronutrient, is considered a beneficial nutritional element. Higher levels of K, Ca, Fe and Zn in the leaves of *Solanum lycopersicum* L. of plants irrigated with magnetically treated water at 250 mT was obtained [89]. Since it exists in an ionized state in plant organs, K is water-soluble and easily dissociable. This explains its mobility and high concentrations in plants. Interestingly, K is considered the main cation in vegetable juices [45]. The K concentration in *Solanum lycopersicum* L. fruits of the SMF plants can be linked to the total protein levels obtained in this group (**Table 3**). The high concentrations of K ions are required for the synthesis of proteins, as well as sugars and starch. Furthermore, K is known to affect the activation of several plant enzymes during the reproductive phase. It is known to be involved in the phase of fruit ripening, in which their demand depends on the needs of fruits that contain plenty of water, such as in *Solanum lycopersicum* L. [12, 90].

In addition, that irrigation of plants of *Phaseolus vulgaris* L. with magnetically treated water increased significantly of levels of K [91]. Also in plants of *Capsicum annuum* L. the irrigation with magnetically treated water with 500 mT, enhance the uptake of K [67]. This mineral is the third most abundant mineral in the human body. This important element for food should have a high concentration with respect to Na. The Na/K ratio measured in this analysis was 0.014 for both experimental groups, which is in agreement with data available in literature [92]. This ratio contributes to the balance of the cellular membrane potential. At the same time, to the validation of these fruits as protective agents against cardiovascular diseases and it have important functions in the muscular and nervous system [92]. Consequently, a diet rich in K is advised, as this nutrient helps to counteract the hypertensive effects of Na. An imbalance in the Na/K ratio cannot only cause hypertension, but can also lead to other diseases [93].

As indicated before, the Ca concentration was significantly increased in fruits of *Solanum lycopersicum* L. irrigated with SMF-treated water (**Table 4**). Ca is the second most abundant chemical element in plants and is a structural component of the cell wall and cell membranes. Furthermore, it is a cofactor of several enzymes involved in metabolic reactions, including ATPases, phosphatases and phospholipases [45]. It regulates the uptake of nitrogen and affects the translocation of carbohydrates and proteins. The exposure of seeds to a MF with an interval of 100 nT to 0.5 mT, had influence in the Ca channels in the cell membrane of *Pisum sativum* [94]. The increased Ca level after irrigation of *Solanum lycopersicum* L. plants with SMF-treated water, suggests that these ions have a sensitive response to the MF. The involvement of Ca in magneto sensitivity and the consequent signal transduction have been a topic of discussion [94]. In this respect, that irrigation with magnetically treated water (3.5–136 mT) increased the Ca concentration in shoots of *Apium graveolens* L. and in the pods of *Lathyrus odoratus*, which is in agreement with the results obtained in this study [95].

In addition, in fruits of *Capsicum chinense* the concentration of Ca, Mg, S, Mn and Zn was increased after the irrigation with MF-treated water in a range between 0 and 156 mT [96]. Ca and K are involved in all stages of plant growth and development, also in responses to environmental changes [90]. On the other hand, the strong antagonism between K with Ca and Mg was described [45]. When the concentration of K increases, Ca and Mg levels decrease as a consequence of a reduced absorption of these minerals [45]. However, the results of the present research do not confirm these data, as both K and Ca concentrations increased in tomato fruits after irrigation with SMF-treated water. In connection to this, *Petroselinum crispum* (Mill) irrigated with magnetically treated water exhibit an increase in K and Ca as compared with plants irrigated with tap water (control) [97]. SMF plants had a K/Ca ratio, which was higher (26.81) than the ratio obtained in the control plants (23.72). Others studies confirmed these effects: an increase of the intracellular Ca, K and Na concentrations were detected in shoots of *Pisum sativum* and *Cicer arietinum* after irrigation with magnetically treated water (3.5 and 136 mT) [98]. In the previous study using the irrigation with magnetically treated water, the K and Ca contents were also increased in *Vicia faba* L. [72].

Concerning P, it plays an important role in photosynthesis, respiration and metabolism. It has an important function in several molecules and cellular structures, such as the diester bonds of nucleic acids and phospholipids [99]. In previous reports, an increase in the concentration of K and P in *Solanum tuberosum* plants was detected when they were irrigated with water treated with an EMF. In contrast to, in the present research no significant effect on the P concentration in tomato fruits was observed. These results coincide with those obtained by others researchers [100]. The concentrations of these minerals remained unchanged in *Solanum lycopersicum* L.

plants irrigated with SMF-treated water were reported. The action of the external MF was beneficial to the formation of mineral crystals in the soil pressure contents of N, P, and other effective nutrients in the soil, thereby further improving soil quality was revealed [101].

With a magnetic funnel (Magnetic Technologies L.L.C. Model No. MFLa, Dubai, U.A.E.) used for water treatment in the cultivation of *Beta vulgaris*, an increase in some chemical elements (N, K, P and Ca) was obtained [102]. In this plant species, the macronutrients (N, P, K, Mg) were increased by irrigation under magnetic water treatments of 0.75 and 1.75 mT [103].

Regarding the Mg concentrations, they were not affected by irrigation with SMF-treated water in *Solanum lycopersicum* L. plants. For the concentrations of micronutrients, only Cu levels were significantly higher in fruits of plants irrigated with SMF-treated water as compared to fruits of control plants (**Table 4**). Cu is essential for growth, involved in many physiological processes, and is part of the active centre of antioxidant enzymes, like superoxide dismutases [104]. Even though it is a micronutrient, it is considered a phytotoxic element in concentrations exceeding 14.7 $mgkg^{-1}$ [90, 105]. However, the Cu concentrations measured in this analysis, did not reach these toxic levels. In *Beta vulgaris* plants irrigated with magnetically treated water with 0.75 and 1.75 mT, an increase in Cu and Zn elements was detected [103].

Other non-essential toxic trace elements such as Cd and Pb were not detected in the fruits of *Solanum lycopersicum* L. The finding of Yusuf et al. [106], supports the results of this research who explained that the magnetic treatment in the irrigation water did not add heavy metals in *Solanum lycopersicum* L. In addition, also other trace elements (Mn, Zn) were analyzed, but fell below the detection limit.

In general, an increased the uptake of nutrients, an enhanced higher crop yield and improved the qualities of fruits were detected. The plants irrigated with SMF-treated water between 20 and 200 mT improved the solubility and absorption of essential minerals, besides their nutritional status. These results confirm that water is probably the primary receptor of the MF in biological systems [107]. It means that different changes in properties of water underexposure to the MF may change the metabolic activity of plants. In this research, it was demonstrated that the chemical and physiological variations caused by MF treatment, increases water absorption, and consequently a higher mineral and nutrient uptake by *Solanum lycopersicum* L. plants. The significance of these results is novel because it can contribute to an increase in the development and growth of plants of this plant species. All of which can lead to an increase crop yields.

There are some hypotheses about the effects of MF on plant growth and development [35, 108, 109]. Although of all the researches, the knowledge of the complex mechanisms involved is of great interest in the novelty of this topic.

5. Conclusions

Based in the results, the irrigation with SMF-treated water (20–200 mT) enhances the accumulation of nutritional elements, especially the content of several essential nutrients in the fruits of *Solanum lycopersicum* L. under greenhouse conditions. An overall increase in the moisture, ashes percentage and total soluble solid levels were obtained. Taken together, these results indicate that irrigation with SMF-treated water improves the nutritional quality of this tomato fruits. The use of MF technology is therefore a useful strategy to improve the nutraceutical value of *Solanum lycopersicum* L., possibly resulting in a positive impact on its antioxidant properties, which should be further investigated in future experiments.

Acknowledgements

The VLIR-IUC UO Project from Belgium and Cuba funded this study. Thanks to the National Center for Applied Electromagnetism (CNEA) of the Universidad of Oriente, in Cuba and the Centre for Environmental Sciences of Hasselt University, in Belgium.

Conflict of interest

The authors declare that they have no conflict of interest.

Author details

Albys Esther Ferrer Dubois[1*], Yilan Fung Boix[1], Clara Martínez Manrique[1], Liliana Gómez Luna[1], Elizabeth Isaac Aleman[1], Sophie Hendrix[2], Natalie Beenaerts[2] and Ann Cuypers[2]

1 National Center of Applied Electromagnetism, University of Oriente (UO), Santiago de Cuba, Cuba

2 Centre for Environmental Sciences, Hasselt University, Hasselt, Belgium

*Address all correspondence to: albysf@gmail.com

References

[1] Khan I, Ahmad S. The impact of natural antioxidants on human health. In: Ahmad S, Al-Shabib NA, editors. Functional Food Products and Sustainable Health. Singapore: Springer; 2020. pp. 11-24. DOI: 10.1007/978-981-15-4716-4_2

[2] Rehman MU, Abdullah KF, Niaz K. Introduction to natural products analysis. In: Sanches Silva A, Nabavi SF, Saeedi M, editors. Recent Advances in Natural Products Analysis. Holland: Elsevier; 2020. pp. 3-15. DOI: 10.1016/B978-0-12-816455-6.00001-9

[3] Salehi B, Sharifi-Rad R, Sharopov F, Namiesnik J, Roointan A, Kamle M, et al. Beneficial effects and potential risks of tomatoes consumption for human health: An overview. Nutrition. 2019;**62**:201-208. DOI: 10.1016/j.nut.2019.01.012

[4] Shrinet K, Singh RK, Chaurasia AK, Tripathi A, Kumar A. Bioactive compounds and their future therapeutic applications. In: Rp S, Häder D-P, editors. Natural Bioactive Compounds: Technological Advancements. Holland: Academic Press Elsevier; 2021. pp. 337-362. DOI: 10.1016/B978-0-12-820655-3.00017-3

[5] Navarro-González I, Periago MJ. El tomate, ¿alimentos aludable y/o funcional? Revista Española de Nutrición Humana y Dietética. 2016;**20**:323-335. DOI: 10.14306/renhyd.20.4.208

[6] Pathak SS, Akhade OP, Bhojane KG, Sadar MD, Folane P, Biyani K. A review on recent techniques of extraction and isolation of lycopene from tomato. International Journal of Research and Review. 2020;7(4):487-490. DOI: 10.4444/ijrr.1002/1927

[7] Pimentel-Moral S, MdlL C-G, Rodríguez-Pérez C, Segura-Carretero A. Recent advances in extraction technologies of phytochemicals applied for the revaluation of agri-food by-products. In: Prakash B, editor. Functional and Preservative Properties of Phytochemicals. Holland: Academic Press Elsevier; 2020. pp. 209-239. DOI: 10.1016/B978-0-12-818593-3.00007-5

[8] Pinto T, Aires A, Cosme F, Bacelar E, Morais MC, Oliveira I, et al. Bioactive (poly) phenols, volatile compounds from vegetables, medicinal and aromatic plants. Food. 2021;**10**:106. DOI: 10.3390/foods10010106

[9] Sariñana-Aldaco O, Sanchez-Chavez E, Fortis-Hernández M, González-FuentesJA,Moreno-ResendezA, Rojas-Duarte A, et al. Improvement of the nutraceutical quality and yield of tomato by application of salicylic acid. Notulae Botanica e Horti Agrobotanici Cluj-Napoca. 2020;**48**:882-892. DOI: 10.15835/nbha48211914

[10] Tan S, Ke Z, Chai D, Miao Y, Luo K, Li W. Lycopene, polyphenols and antioxidant activities of three characteristic tomato cultivars subjected to two drying methods. Food Chemistry. 2021;**338**:128062. DOI: 10.1016/j.foodchem.2020.128062

[11] Ahmad N. Nutraceutical and bioactive healthy compounds in vegetable crops. IJCS. 2020;**8**:3078-3086. DOI: 10.22271/chemi.2020.v8.i4ak.10120

[12] Larín, M., Díaz, L., y Serrano, R. Cultivo de tomate (*Lycopersicon esculentum*). Salvador. CENTA. 2018. Available from: http://centa.gob.sv/

docs/guias/hortalizas/Guia%20Centa_Tomate%202019.pdf

[13] Anwar R, Fatima T, Mattoo A. Tomatoes: A model crop of *solanaceous* plants. In: Oxford Research Encyclopedia of Environmental Science. New York, United States: Oxford University Press; 2019. pp. 1-50. DOI: 10.1093/acrefore/9780199389414.013.223

[14] Hiwasa-Tanase K. Fruit ripening in tomato and its modification by molecular breeding techniques. In: Ezura H, Ariizumi T, Garcia-Mas J, Rose J, editors. Functional Genomics and Biotechnology in *Solanaceae* and *Cucurbitaceae* Crops. Biotechnology in Agriculture and Forestry. Vol. 70. Berlin, Heidelberg: Springer; 2016. DOI: 10.1007/978-3-662-48535-4_10

[15] Umhsapiens. El gusto por la genética en una fruta de tomate. UMH SAPIENS. Universidad Miguel Hernández (UMH). Elche. España.[Internet]. 2014. Available from: https://umhsapiens.com/el-gusto-por-la-genetica-en-un-fruto-de-tomate/

[16] Wang A, Chen D, Ma Q, Rose JKC, Fei Z, Liu Y, et al. The tomato high pigment1/damaged DNA binding protein 1 gene contributes to regulation of fruit ripening. Horticulture Research. 2019;**6**:1-10. DOI: 10.1038/s41438-018-0093-3

[17] Aguilera JG, Martín RM. Água tratada magneticamente estimula a germinação e desenvolvimento de mudas de *Solanum lycopersicum* L. Revista Brasileira de Agropecuária Sustentável. 2016;**6**:47-53. DOI: 10.21206/rbas.v6il.320

[18] Almeida RC, Deus FP, Diotto AV, Teixeira ALP, Lemos LTO, Pizetta SC, et al. Desenvolvimento vegetativo do tomateiro sweetheaven irrigado com água tratada magneticamente. Brazilian Journal of Development. 2020;**6**:5428-5434. DOI: 10.34117/bjdv6n2-007

[19] De Souza A, Sueiro L, García D, Porras E. Extremely low frequency non-uniform magnetic fields improve tomato seed germination and early seedling growth. Seed Science and Technology. 2010;**38**:61-72. DOI: 10.15258/sst.2010.38.1.06

[20] Dubois AF, Lamezon Y, Fung Y, Isaac E. Aplicación de campo magnético para mejorar la calidad en plantas de tomate (*Lycopersicon esculentum*). Revista Tecnología Química Edición Especial. 2007:-85. DOI: 81. Available from: https://magnetismocnea.files.wordpress.com/2010/08/agricultura.pdf

[21] El-Shafik El-Zawily A, Meleha M, El-Sawy M, El-Attar E-H, Bayoumi Y, Alshaal T. Application of magnetic field improves growth, yield and fruit quality of tomato irrigated alternatively by fresh and agricultural drainage water. Ecotoxicology and Environmental Safety. 2019;**181**:248-254. DOI: 10.1016/j.ecoenv.2019.06.018

[22] Fakhri N, Mehdaoui HY, Elloumi N, Kallel M. Magnetic Treatment Effects on Salt Water and Tomato Plants rowth. New York, United States: Springer International Publishing; 2018. DOI: 10.1007/978-3-319-70548-4_316

[23] Jedlička J, Paulen O, Ailer Š. Influence of magnetic field on germination, growth and production of tomato. Potravinarstvo Slovak Journal of Food Sciences. 2014;**8**:150-154. DOI: 10.5219/349

[24] Ogunlela A, Yusuf K. Effects of water stress on growth and yield of tomato irrigated with magnetically treated water. Nigerian Journal of Pure and Applied Sciences. 2016;**29**:2834-2835. DOI: 10.19240/njpas.2016.A21

[25] Torres C, Díaz JE, Cabal PA. Efecto de campos magnéticos en la germinación de semillas de arroz (*Oryza sativa* L.) y tomate (*Solanum lycopersicum* L.), Agronomía Colombiana. 2008;**26**:177-185. ISSN 0120-9965. Available from: http://www.scielo.org.co/scielo.php?script=sci_arttext&pid=S0120-99652008000200002&lng=en&nrm=iso

[26] Yusuf KO, Adeleye AA. Effect of magnetic treatment of irrigation water on growth and yield of maize under different water deficit conditions. FUOYE. Journal of Engineering and Technology. 2018;**3**:83-87. DOI: 10.46792/fuoyejet.v3i2.193

[27] Dubois AEF, Aguilera JG, Boix YF, Alemán EI, Zuffo AM. Use of GREMAG® technology to improve seed germination and seedling survival. In: Zuffo AM, Aguilera JG de Oliveira BR, editors. Ciência em Foco. Pantanal, Brasil. 2019. p. 138-149. DOI: 10.22533/at.ed.567181510

[28] Fadil K, Merza TK. Physiological study of irrigation with treated magnetic water on some of vegetative characteristics and yield of three wheat cultivars in Najaf. Kufa Journal for Agriculture Sciences. 2014;**6**:140-169. Available from: https://www.iasj.net/iasj/article/110820

[29] Oduardo DZ, Fernández PR, Dubois AF, Boix YF, Aleman EI, Kindelan GA. Producción de tomate (*Solanum lycopersicum* L.) bajo riego con agua magnetizada en casa de cultivo protegido. Ciencia en su PC. 2020;**1**:60-74. Available from: https://www.redalyc.org/journal/1813/181363107005/html/

[30] De Souza-Torres A, Sueiro-Pelegrın L, Zambrano-Reyes M, Macias-Socarras I, Gonzalez-Posada M, Garcıa-Fernandez D. Extremely low frequency non-uniform magnetic fields induce changes in water relations, photosynthesis and tomato plant growth. International Journal of Radiation Biology. 2020;**96**:951-957. DOI: 10.1080/09553002.2020.1748912

[31] Dhawi F. Why magnetic fields are used to enhance a plant's growth and productivity? Annual Research & Review in Biology. 2014;**4**:886. DOI: 10.9734/ARRB/2014/5983

[32] Duarte CE, Sarmiento O. Influencia de tratamiento alternativo del agua de riego en los requerimientos hídricos de algunas hortalizas para condiciones de deficiencias hídricas. Ingeniería de Recursos Naturales y del Ambiente. 2013;**1**:91-99. Available from: https://www.redalyc.org/articulo.oa?id=231130851009

[33] Guruprasad K, Shine M, Joshi J. Impact of Magnetic Field on Crop Plants. Breeding and Genetic Engineering. The Biology and Biotechnology Research. 2016. Available from: https://www.researchgate.net/publication/322277518

[34] Maffei ME. Plant responses to electromagnetic fields. In: Biological and Medical Aspects of Electromagnetic Fields. 4th ed. Boca Raton, Florida, United States: CRC Press; 2018. pp. 89-110. eBook ISBN 9781315186641. Available from: https://www.crcpress.com/Biological-and-Medical-Aspects-of-Electromagnetic-Fields-Fourth-Edition/Greenebaum-Barnes/p/book/9781138735262

[35] Sarraf M, Kataria S, Taimourya H, Santos LO, Menegatti RD, Jain M, et al. Magnetic field applications in plants: An overview. Plants. 2020;**9**(9):1139. DOI: 10.3390/plants9091139

[36] Teixeira J, Dobránszki J. Magnetic fields: How is plant growth and

development impacted? Protoplasma. 2016;**253**:1-18. DOI: 10.1007/s00709-015-0820-7

[37] Vian A, Davies E, Gendraud M, Bonnet P. Plant responses to high frequency electromagnetic fields. BioMed Research International. 2016;**2016**:1-14. DOI: 10.1155/2016/1830262

[38] Dubois AEF, Leite GO, Rocha JBT. Irrigation of *Solanum lycopersicum* L. with magnetically treated water increases antioxidant properties of its tomato fruits. Electromagnetic Biology and Medicine. 2013;**32**:355-362. DOI: 10.3109/15368378.2012.721847

[39] Dubois AF, Boix YF, Leite GO, Luna LG, Boligon AA, Athayde ML, et al. Evaluación de la irrigación con el uso de campos magnéticos estáticos sobre la síntesis de metabolitos secundarios en *Solanum lycopersicum* L. (tomate). Investigación y Saberes. 2012;**1**:54-61. Available from: http://revistasdigitales.utelvt.edu.ec/revista/index.php/investigacion_y_saberes/article/view/19

[40] González-Casado S, Martín-Belloso O, Elez-Martínez P, Soliva-Fortuny R. Enhancing the carotenoid content of tomato fruit with pulsed electric field treatments: Effects on respiratory activity and quality attributes. Postharvest Biology and Technology. 2018;**137**:113-118. DOI: 10.1016/j.postharvbio.2017.11.017

[41] Luengo E, Álvarez I, Raso J. Improving carotenoid extraction from tomato waste by pulsed electric fields. Frontiers in Nutrition. 2014;**1**:12. DOI: 10.3389/fnut.2014.00012

[42] Vallverdú A, Oms G, Odriozola I, Lamuela R, Martín O, Elez P. Effects of pulsed electric fields on the bioactive compound content and antioxidant capacity of tomato fruit. Journal of Agricultural and Food Chemistry. 2012;**60**:3126-3134. DOI: 10.1021/jf205216m

[43] Ferreira S, Freitas R, Lazzari E. Padrão de identidade e qualidade do tomate (*Lycopersicon esculentum* Mill.) de mesa. Ciência Rural. 2004;**34**:329-335. DOI: 10.1590/S0103-84782004000100054

[44] Casanova A, Gómez O, Depestre T, Igarza A, Leon M, Santos R, Chailloux M. Manual para la producción protegida de hortalizas. 2da. Ed. Inst. Inv. Hortícolas “Liliana Dimitrova” . MINAG. 2007. La Habana, Cuba. 138 p. ISBN: 9597111373, 9789597111375

[45] Navarro G, Navarro S. Química Agrícola. Química del Suelo y de los nutrientes esenciales para las plantas. 3ra ed. España: Ed. Mundi-Prensa; 2013. 508 p. ISBN: 9788484766568, 848476656X

[46] Mayea S. Microbiología Agropecuaria. 2d ed. La Habana, Cuba: Editorial Félix Varela; 2004

[47] Gilart F, Deas D, Ferrer D, López P, Ribeaux G, Castillo J. High flow capacity devices for anti-scale magnetic treatment of water. Chemical Engineering and Processing: Process Intensification. 2013;**70**:211-216. DOI: 10.1016/j.cep.2013.04.001

[48] AOAC. Official Methods of Analysis. 15th ed. Washington, D.C.: Association of Official Analytical Chemists; 1990

[49] Park YW. Determination of moisture and ash contents of food. In: Nollet LML, editor. Handbook of Food Analysis. New York, USA: Marcel Dekker Inc; 2004. pp. 55-82

[50] AOAC. Official Methods of Analysis. 14th ed. Washington, D.C.: Assoc. Offic. Anal. Chem.; 1984

[51] Miranda M, Cuellar A. In: Varela F, editor. Manual de prácticas de laboratorio Farmacognosia y Productos Naturales. Cuba: La Habana; 2000. 34-49 p. ISBN: 978-959-07-1795-6

[52] Linstrom PJ, Mallard WG. NIST Chemistry Webbook, NIST Standard Reference 69. NGaithersburg: National Institute of Standards and Technology; 2005

[53] Hasaani AS, Hadi ZL, Hasaani AS, Rasheed KA. Experimental study of the interaction of magnetic fields with flowing water. International Journal of Basic and Applied Sciences. 2015;**3**(3):1-8. Available from: https://www.academia.edu/32188601/Experimental_Study_of_the_Interaction_of_Magnetic_Fields_with_Flowing_Water:

[54] Yadollahpour A, Rashidi S, Fatatemeh K. Applications of magnetic water technology in farming and agriculture development: A review of recent advances. Current World Environment: An International Research Journal of Environmental Sciences. 2014;**9**:695-703. DOI: 10.12944/CWE.9.3.56

[55] Julca-Otiniano A, Meneses-Florián L, Blas-Sevillano R, Bello-Amez S. La materia orgánica, importancia y experiencia de su uso en la agricultura. Idesia (Arica). 2006;**24**:49-61. DOI: 10.4067/S0718-34292006000100009

[56] Esmaeilnezhad E, Choi HJ, Schaffie M, Gholizadeh M, Ranjbar M. Characteristics and applications of magnetized water as a green technology. Journal of Cleaner Production. 2017;**161**:908-921. DOI: 10.1016/j.jclepro.2017.05.166

[57] Holysz L, Szczes A, Chibowski E. Effects of a static magnetic field on water and electrolyte solutions. Journal of Colloid and Interface Science. 2007;**316**:996-1002. DOI: 10.1016/j.jcis.2007.08.026

[58] Martínez CM, Machado R, Casals M. Evaluación de los efectos del tratamiento magnético a la solución CM-95 por medidas de conductividad eléctrica. Revista Cubana de Química. 2011;**23**:76-79. ISSN: 0258-5995. Available from: https://www.redalyc.org/pdf/4435/443543722010.pdf

[59] Pang XF, Shen G-F. The changes of physical properties of water arising from the magnetic field and its mechanism. Modern Physics Letters B. 2013;**27**:1-9. DOI: 10.1142/S021798491350228X

[60] Mohamed AI, Ebead BM. Effect of magnetic treated irrigation water on salt removal from a sandy soil and on the availability of certain nutrients. International Journal of Engineering. 2013;**2**:36-44. ISSN 2305-8269. Available from: http://magtech.com.pk/pdf/agriculture-7.pdf

[61] Abedinpour M, Rohani E. Effects of magnetized water application on soil and maize growth indices under different amounts of salt in the water. Journal of Water Reuse and Desalination. 2017;**7**:319-325. DOI: 10.2166/wrd.2016.216

[62] El-Dayeim DE-TB. Evaluation of magnetic technology for vegetable production under drip irrigation system [doctoral dissertation]. Sudan: University of Khartoum; 2015. Available from: http://www.secheresse.info/spip.php?article53125

[63] Shahin MM, Mashhour AMA, Abd-Elhady ESE. Effect of magnetized irrigation water and seeds on some water properties, growth parameter and yield productivity of cucumber plant. Current Science International.

2016;**5**:152-164. ISSN: 2077-4435. Available from: https://www.doa.go.th/seed/wp-content/uploads/2021/08/vegetable12.pdf

[64] Pang XF, Zhong LS. The suspension of water using a superconductive magnetic-field and its features. IEEE Transactions on Applied Superconductivity. 2016;**26**(7):1-4. DOI: 10.1109/TASC.2016.2603598

[65] Pang XF. The experimental evidences of the magnetism of water by magnetic-field treatment applied superconductivity. IEEE Transactions on Applied Superconductivity. 2014;**24**(5):1-6. DOI: 10.1109/TASC.2014.2340455

[66] Pang X-F, Deng B. The changes of macroscopic features and microscopic structures of water under influence of magnetic field. Physica B: Condensed Matter. 2008;**403**:3571-3577. DOI: 10.1016/j.physb.2008.05.032

[67] Rawabdeh H, Shiyab S, Shibli R. The effect of irrigation by magnetically water on chlorophyll and macroelements uptake of pepper (*Capsicum annuum* L.) Jordan. The Journal of Agricultural Science. 2014;**10**:205-214. Available from: https://journals.ju.edu.jo/JJAS/article/viewFile/6757/3897

[68] Mghaiouini R, Elaouad A, Taimoury H, Sabir I, Chibi F, Hozayn M, et al. Influence of the electromagnetic device aqua 4d on water quality and germination of lettuce (*Lactuca sativa* L.). International Journal of Current Engineering and Technology. 2020;**10**:19-24. DOI: 10.14741/ijcet/v.10.1.4

[69] Aly MA, Thanaa ME, Os M, Maze AAA. Effect of magnetic irrigation water and some anti-salinity substances on the growth and production of Valencia orange. Middle East Journal of Agriculture Research. 2015;**4**:88-89. ISSN: 2077-4605. Available from: https://www.curresweb.com/mejar/mejar/2015/88-98.pdf

[70] Pang X-F, Deng B, Tang B. Influences of magnetic field on macroscopic properties of water. Modern Physics Letters B. 2012;**26**:1-13. DOI: 10.1142/S0217984912500698

[71] Berni R, Luyckx M, Xu X, Legay S, Sergeant K, Hausman JF, et al. Reactive oxygen species and heavy metal stress in plants: Impact on the cell wall and secondary metabolism. Environmental and Experimental Botany. 2019;**161**:98-106. DOI: 10.1016/j.envexpbot.2018.10.017

[72] El Sayed HESA. Impact of magnetic water irrigation for improve the growth, chemical composition and yield production of broad bean (*Vicia faba* L.) plant. American Journal of Experimental Agriculture. 2014;**4**:476-496. DOI: 10.9734/AJEA/2014/7468

[73] Hozayn M, Abdel-Monem AA, Qados AMSA. Irrigation with magnetized water: A novel tool for improving crop production in Egypt. In: Fifteenth International Water Technology Conference, IWTC-15; 2011; Alexandria, Egypt. pp. 4206-4222. Available from: http://www.magtech.com.pk/pdf/agriculture-13.pdf

[74] Socorro A, García F. Simulation of magnetic field effect on a seed embryo cell. International Agrophysics. 2012;**26**:167-173. DOI: 10.2478/v10247-012-0024-8

[75] Poinapen D, Brown DCW, Beeharry GK. Seed orientation and magnetic field strength have more influence on tomato seed performance than relative humidity and duration of exposure to non-uniform static magnetic fields. Journal of Plant Physiology.

2013;**170**:1251-1258. DOI: 10.1016/j.jplph.2013.04.016

[76] Shahzad T, Ahmad I, Choudhry S, Saeed M, Khan M. DPPH free radical scavenging activity of tomato, cherry tomato and watermelon: Lycopene extraction, purification and quantification. International Journal of Pharmacy and Pharmaceutical Sciences. 2014;**6**:223-228. ISSN: 2656-0097. EISSN: 0975-1491. Available from: https://hero.epa.gov/hero/index.cfm/reference/details/reference_id/8222123

[77] Canti MG. Aspects of the chemical and microscopic characteristics of plant ashes found in archaeological soils. Catena. 2003;**54**:339-361. DOI: 10.1016/S0341-8162 (03)00127-9

[78] Abdul Qados A, Hozayn M. Magnetic water technology, a novel tool to increase growth, yield and chemical constituents of lentil under greenhouse condition. American-Eurasian Journal of Agricultural & Environmental Sciences. 2010;**7**:457-462. Available from: http://www.idosi.org/aejaes/jaes7(4)/13.pdf

[79] Sadeghipour O, Aghaei P. Improving the growth of cowpea (*Vigna unguiculata* L. Walp.) by magnetized water. Journal of Biodiversity and Environmental Sciences. 2013;**3**:37-43. ISSN: 2220-6663. Available from: http://www.innspub.net/wp-content/uploads/file/JBES-V3No1-p37-43.pdf

[80] Guil J, Guerrero L, Rebolloso M, Fuentes M. Nutrient composition and antioxidant activity of eight tomato (*Lycopersicon esculentum*) varieties. Journal of Food Composition and Analysis. 2009;**22**:123-129. DOI: 0.1016/j.jfca.2008.10.012

[81] Pinela J, Barros L, Carvalho A, Ferreira I. Nutritional composition and antioxidant activity of four tomato (*Lycopersicon esculentum* L.) farmer' varieties in Northeastern Portugal homegardens. Food and Chemical Toxicology. 2012;**50**:829-834. DOI: 10.1016/j.fct.2011.11.045

[82] Albuquerque W, Costa M, Fernandes T, Porto A. Evidences of the static magnetic field influence on cellular systems. Progress in Biophysics and Molecular Biology. 2016;**121**:16-28. DOI: 10.1016/j.pbiomolbio.2016.03.003

[83] Hernández-Yépez J-N, De La Haba M-J, Sánchez M-T. Effect of different prepackaging treatments on the physical/chemical quality of margariteño tomatoes during postharvest storage at room temperature. Journal of Food Quality. 2013;**36**:113-120. DOI: 10.1111/jfq.12022

[84] Panjai L. The role of post-harvest supplementary light exposure for ripening and quality development of tomato fruit [dissertation]. Online-Ausgabe: Rheinische Friedrich-Wilhelms-Universität Bonn; 2020. Available from: https://nbn-resolving.org/urn:nbn:de:hbz:5-60488

[85] Križaj D, Valenčič V. The effect of ELF magnetic fields and temperature on differential plant growth. Journal of Bioelectricity. 1989;**8**:159-165. DOI: 10.3109/15368378909020954

[86] Yusuf KO, Ogunbamowo TR, Obalowu RO. Effect of magnetized water on water use efficiency, yield and nutritional qualities of watermelon under deficit irrigation. Agricultural Engineering International. CIGR Journal. 2020;**22**:51-60. Available from: https://cigrjournal.org/index.php/Ejounral/article/view/5833/3363

[87] Dhawi F. Why magnetic fields are used to enhance a plant's growth

and productivity? Annual Research & Review in Biology. 2014;**4**:886-896. DOI: 10.9734/ARRB/2014/5983

[88] Anjum S, Hamid A, Ghafoor A, Tahira R, Shah SZA, Awan SI, et al. Evaluation of biochemical potential in tomato (*Solanum lycopersicum*) germplasms. Pakistan Journal of Agricultural Sciences. 2020;**57**:177-187. DOI: 10.21162/PAKJAS/20.8140

[89] El Shokali AA, Abdelbagi AM, Abdallah M. Enhancing on the mineral elements of exposure to magnetic field in plants leaves. Journal of Basic and Applied Sciences. 2015;**11**:440-444. ISSN: 1814-8085. Available from: http://www.lifescienceglobal.con/inde

[90] Barker AV, Pilbeam DJ. Handbook of Plant Nutrition. 2nd ed. Boca Raton, Florida, USA: CRC Press.Taylor and Francis Group; 2015. 773 p. DOI: 10.1201/b18458

[91] Khattab E, Mona M, Hegazi A, Arafa A. Magnetic treatment of irrigation water for improving vegetative growth, fresh and dry yield of bean (*Phaseolus vulgaris* L.) grown under plastic house. Egyptian Journal of Agricultural Research. 2014;**92**:1395-1411. DOI: 10.21608/ejar.2014.156788

[92] Elbadrawy E, Sello A. Evaluation of nutritional value and antioxidant activity of tomato peel extracts. Arabian Journal of Chemistry. 2011;**11**:1-9. DOI: 10.1016/j.arabjc.2011.11.011

[93] Houston MC. Treatment of hypertension with nutrition and nutraceutical supplements: Part 1. Alternative and Complementary Therapies. 2018;**24**:260-275. DOI: 10.1089/act.2018.29191.mho

[94] Belyavskaya NA. Biological effects due to weak magnetic field on plants. Advances in Space Research. 2004;**34**:1566-1574. DOI: 10.1016/j.asr.2004.01.021

[95] Maheshwari B, Grewal H. Magnetic treatment of irrigation water: Its effects on vegetable crop yield and water productivity. Agricultural Water Management. 2009;**96**:1229-1236. DOI: 10.1016/j.agwat.2009.03.016

[96] Ospina-Salazar DI, Benavides-Bolaños JA, Zuñiga-Escobar O, Muñoz-Perea CG. Photosynthesis and biomass yield an in tabasco pepper, radish and maize subjected to magnetically treated water. Corpoca Cienc. Tecnol. Agrepecuaria. 2018;**19**:201-305. DOI: 10.21930/rcta.vol19_num2_art:537

[97] Mohamed HH, Elnasri M. The impact of magnetized water on some physiological characters of parsley *Petroselinum crispum* (Mill). International Journal of Innovative Science and Research Technology. 2019;**4**:725-729. ISSN No:-2456-2165. Available: https://ijisrt.com/wp-content/uploads/2019/05/IJISRT19MA205.pdf

[98] Grewal HS, Maheshwari BL. Magnetic treatment of irrigation water and snow pea and chickpea seeds enhances early growth and nutrient contents of seedlings. Bioelectromagnetics. 2011;**32**:58-65. DOI: 10.1002/bem.20615

[99] Azcón-Bieto J, Talón M. Fundamentos de Fisiología Vegetal. España: McGraw Hill Interamericana; 2008. 134 p

[100] García G. Revisión ingesta de nutrientes: Conceptos y recomendaciones internacionales (1ª Parte). Nutrición Hospitalaria. 2006;**21**:291-299 Available from: https://scielo.isciii.es/pdf/nh/v21n3/revision.pdf

[101] Cui H, Liu X, Jing R, Zhang M, Wang L, Zheng L, et al. Irrigation with magnetized water affects the soil microenvironment and fruit quality of eggplants in a covered vegetable production system in Shouguang City, China. Journal of Soil Science and Plant Nutrition. 2020;**20**:2684-2697. DOI: 10.1007/s42729-020-00334-7

[102] Ibrahim MO, Mutwali EM, Salih EA, Ismail WH. Effect of magnetized water on the germination and some yield components of swiss chard (*Beta vulgaris* var. cicla). International Journal of Innovative Science and Research Technology. 2018;**3**:410-415. ISSN:-2456-2165. Available from: https://ijisrt.com/wp-content/uploads/2018/10/IJISRT18SP217.pdf

[103] Hozayn M, El-Dahshouri M, Salama AM, Abdelraouf R, Ahmed A. Influence of magnetic brackish-water treatments on growth, anatomical structure, yield and quality of sugar beet (*Beta vulgaris* L.). Plant Archives. 2020;**20**:8271-8278. e-ISSN:2581-6063 (online), ISSN:0972-5210. Available from: http://www.plantarchives.org/20-2/8271-8278%20(7091).pdf

[104] Muñoz VO, Casas CP, Vázquez SM, Espinal MJM. Los procesos redox y su relevancia en la vida. In: Muñoz Ocotero V, editor. Lecturas de apoyo para comprender mejor la Química. México: Facultad de Ciencias, UNAM; 2015. pp. 105-114. Available from: http://librosoa.unam.mx/bitstream/handle/123456789/267/30.pdf?sequence=2&isAllowed=y#page=103

[105] Bressy FC, Geysa BB, Barbosa IS, Teixeira LSG, Korn MGA. Determination of trace element concentrations in tomato samples at different stages of maturation by ICP OES and ICP-MS following microwave-assisted digestion. Microchemical Journal. 2013;**109**:145-149. DOI: 10.1016/j.microc.2012.03.010

[106] Yusuf KO, Baiyeri MR, Sakariyah SA. Effects of magnetically-treated water on tomato yield and uptake of heavy metals under water deficit conditions. In: 8th International Conference on Agricultural, Environment, Biology and Medical Sciences of International Institute of Chemical, Biological and Environmental Engineering (IICBEE); 2017; Dubai. 2017. pp. 11-20. ISBN 978-93-86878-07-6. Available from: http://heaig.org/images/proceedings_pdf/C1217203.pdf

[107] Shalatonin V. Effect of unipolar magnetic field on macroscopic properties of distilled water international. Journal of Pharma Medicine and Biological Sciences. 2017;**6**:89-93. DOI: 10.18178/ijpmbs.6.3.89-93

[108] Zhang X, Yarema K, Xu A. Impact of static magnetic field (SMF) on microorganisms, plants and animals. In: Biological Effects of Static Magnetic Fields. Singapore: Springer; 2017. pp. 133-172. DOI: 10.1007/978-981-10-3579-1_5

[109] Rifna EJ, Ratish Ramanan K, Mahendran R. Emerging technology applications for improving seed germination. Trends in Food Science & Technology. 2019;**86**:95-108. DOI: 10.1016/j.tifs.2019.02.029